# PLUMBERS AND PIPE FITTERS LIBRARY
## Volume II

### Heating • Welding • Air Conditioning

by Charles McConnell
revised by Tom Philbin

Macmillan Publishing Company
New York

Collier Macmillan Publishers
London

THIRD EDITION
THIRD PRINTING          FEB 0 8 1988

Macmillan Publishing Company
866 Third Avenue, New York, N.Y. 10022
Collier Macmillan Canada, Inc.

Library of Congress Cataloging-in-Publication Data

McConnell, Charles.
    Plumbers and pipe fitters library.

    Includes indexes.
    Contents: v. 1. Materials, tools, roughing-in—
v. 2. Welding, heating, air conditioning—v. 3.
Water supply, drainage, calculations.
    1. Plumbing—Handbooks, manuals, etc.   2. Pipe-
fitting—Handbooks, manuals, etc.   I. Philbin, Tom,
1934-     .   II. Title.   III. Title: Plumbers and
pipefitters library.
[TH6125.M4   1986]        696'.1        86-8743
ISBN 0-672-23386-X (v. 2)

Macmillan books are available at special discounts for bulk purchases
for sales promotions, premiums, fund-raising, or educational use.
For details, contact:

        Special Sales Director
        Macmillan Publishing Company
        866 Third Avenue
        New York, N.Y. 10022

10 9 8 7 6 5 4 3

Printed in the United States of America

# Foreword

Plumbing and pipe fitting play major roles in the construction of residential, commercial, and industrial buildings. Of all the building trades, the plumbing trade is most essential to the health and well-being of the community in general, and to the occupants of buildings in particular. It is the responsibility of each and every person doing any work to uphold the vital trust placed in him or her.

Each plumbing installation is governed by the rules and regulations set forth in the plumbing codes adopted from standards established at a local, state, or federal level. In addition, each installation is subject to inspection by a licensed inspector to ensure compliance with all rules and regulations.

The books in this series have been written to aid people who wish to become plumbers as well as those already on the job. The book has also been designed to help do-it-yourselfers.

This, the second of three volumes, deals with heating systems in which plumbing plays a part, air-conditioning systems, and brazing and welding. Heating and air conditioning are natural companions of the plumbing industry and, as such, have been adopted by many plumbers as a part of their normal business. Welding and brazing are being used extensively in plumbing and pipe-fitting installations and should therefore be of concern to nearly every person in the plumbing trade.

# Contents

## CHAPTER 1

## CHAPTER 2

## CHAPTER 3

# CHAPTER 4

# CHAPTER 5

# CHAPTER 6

# CHAPTER 7

# CHAPTER 8

# CHAPTER 9

# CHAPTER 10

# Brazing and Welding

Brazing and welding are used extensively in the plumbing and pipe-fitting trades. Brazing has taken the place of many operations that formerly were performed by soldering. The brazing process is usually simpler and faster, and results in stronger joints. Welding is employed in many installations, especially where larger-diameter pipes are concerned. Installation time and costs can be materially reduced by using this method of joining pipe and fittings. Every plumber and pipe fitter should be familiar with the basic principles and techniques of brazing and welding.

## BRAZING
### (Courtesy NCG, Division of Chemetron Corp.)

Brazing operations are specifically different from those applicable to fusion welding. In brazing, the brazing alloy is applied at a temperature *below* the melting temperature of the metal being brazed. In brazing a joint, the procedure is an adhesion of a brazing alloy to the surface and into the porous structure of the metal that is in the process of being brazed. In almost all cases, a brazed joint is at least as strong as the brazed metal.

## Nonferrous Metals

Nonferrous metals such as copper and copper-zinc braze well with a low-temperature brazing alloy or a silver alloy. Copper and copper-zinc metals have a melting temperature of approximately 1670°F. The melting point of the brazing alloy must be below the melting temperature of the metal.

Aluminum and aluminum-alloy brazing requires some experience by the operator. Aluminum brazing alloy melts at approximately 200°F, which is under the melting temperature of aluminum in its pure state. The brazing of aluminum alloys is difficult because the differential in melting temperatures is under 200°F.

## Ferrous Metals

In brazing most ferrous metals, the bonding temperatures must not exceed 1400°F, which is below the critical temperature of these metals. This critical point should not be exceeded for the following reasons:

**Cast Iron.** Cast iron contains a high percentage of silicon and carbon. A majority of the carbon content is in graphitic form that will, when heated to temperatures in excess of 1400°F, flow to the outer surface, preventing the brazing alloy from penetrating the porous structure and making it difficult to obtain good surface adhesion. Silicon has the same effect.

**Malleable Iron.** Malleable iron is a heat-treated form of cast iron, and when heated above the critical zone (1575°F), the structure returns to that of cast iron. When this occurs, the metal is no longer ductile and will not be resistant to shock.

**Steel.** When heated to a temperature above 1500°F, steel will alloy with the copper in the brazing material and become very brittle. Copper, when properly alloyed with steel, produces a high tensile strength, but when alloyed by brazing, the temperature is too uneven to obtain uniform distribution of the copper in the steel, thereby creating not strength, but weakness.

## Stainless Steel

Stainless steels are readily brazed with silver alloys, but are difficult to braze with copper-zinc alloys. While the brazing of stainless steels is successful, there is an element of danger that should be considered before brazing is performed. That is, when stainless steel is heated slowly and cooled slowly, there is a great possibility of forming carbides at the grain boundary, and carbide precipitates are very brittle. Therefore, stainless steels (unless stabilized) should not be brazed if the joints are to be subjected to shock or vibrating stress.

## Brazing Flux

Flux is used in all brazing operations except with copper-to-copper joints. Do not rely on the chemical action of the flux alone to properly clean slate, dirt, and foreign matter from the metal. Follow the instructions on the flux container for good-quality brazing.

## Joint Preparation

Before attempting brazing, be sure that all joints are clean and properly prepared. Joints should have close fits and be uniform in shape. The use of excessive amounts of brazing alloys is expensive; it will result in excessive warpage and stress in some operations, and often results in low physical properties.

## Brazing With Silver Alloys

Silver brazing alloys are available in a variety of silver contents, and almost all metals (except magnesium and aluminum) can be brazed with one of the alloys. Certain procedures must be followed in order to obtain quality results.

A high capillary attraction to metals is provided by silver alloys. As a result, a close clearance between metal parts is necessary for a good joint. Specifically, .002″ to .003″ clearance makes a stronger

9

joint than do larger clearances, which require a greater amount of alloy. Not only are proper fits necessary, but clean joints are a must (fingerprints from handling will often cause poor adhesion and an inferior joint). Use an emery cloth to ensure that they are clean.

After final cleaning, the parts are fluxed with a brush and assembled. Certain types of joints may be made by inserting shims or ring stock in or on the joint after being fluxed. A tip of ample size to produce a soft flame should be used for the heating, applying the heat evenly to the entire joint to avoid distortion by overheating some parts and underheating others. When the flux changes to a liquid state, the alloy will flow freely to all parts of the joint.

## SOCKET BRAZING
### (Courtesy of Kennedy Valve Mfg. Co., Inc.)

Some piping and valves are provided with brazing sockets designed to permit a pipe-to-pipe or pipe-to-valve connection by means of a silver brazed joint. These joints have high tensile strength and will hold firm at any temperature to which the valve, pipe, or tubing can be safely subjected.

Silver-brazed joints resist corrosion since the brazing alloy fuses with the bronze of the valve and the material of the pipe or tubing. This results in a bond of greater strength and corrosion resistance than either the valve or the tubing.

Brazing socket valves (Fig. 1-1) have threadless, smooth socket ends, precision machined to close tolerances, into which standard pipe-sized outside-diameter brass or copper pipe or tubing is fitted. At the inner end of the socket, a square shoulder acts as a stop to limit pipe insertion.

The accurately controlled clearance between the pipe and valve socket assures the capillary action of the molten brazing alloy. When the pipe and end of the valve are properly heated, the brazing alloy is drawn into the socket by this capillary action, flowing to the full depth and between the shoulder end of the pipe. This bonds the entire contacting surfaces of the pipe and valve socket.

Valves of this type are generally suitable for use in any nonferrous piping installation in industrial plants, power plants, oil refineries, chemical plants, hospitals, ships, and so forth. They are also suitable for use in systems handling oil, hot or cold water, compressed air, acid, Freon, carbon dioxide, and many other liquids and gases.

**Fig. 1-1. Standard bronze gate valve with brazing sockets.**

Courtesy Kennedy Valve Mfg. Co.

Cleanliness is the first requirement in the installation of socket brazing valves. The connecting pipe or tubing must be cut squarely, reamed, burred, and polished with emery cloth. The valve sockets and the faces of the hex ends must be wiped clean and polished with emery cloth. The polished surfaces should be coated with flux and the pipe inserted in the socket until it butts against the shoulder. See Fig. 1-2.

Open the valve and apply a flame to the pipe about 1″ from the valve and directed away from the valve. Move the flame around the pipe until the flux changes to beads (Fig. 1-3). Then move the flame to the valve hex, but still directed toward the pipe. When the condition of the flux indicates that the valve end is heated sufficiently, apply fluxed brazing alloy to the joint (Fig 1-4). Sweep the

11

flame across the valve hex and pipe while the molten brazing alloy is drawn into the socket.

Remove the flame and allow the joint to cool. If the joint has been properly made, there will be an unbroken fillet of brazing alloy completely around the edge of the socket.

A low-silver brazing alloy is recommended for this type of brazing. This alloy contains only about 15% silver, will start to melt at 1185°F, and will be completely liquid and free flowing at 1300°F.

FLUX

Courtesy Kennedy Valve Mfg. Co.

Fig. 1-2. Flux is applied to pipe end, hex and face, and socket.

## OXYACETYLENE WELDING

Oxyacetylene welding is a process of joining metals by heating to the melting point the surface to be joined and then fusing together the molten metal. The heat used for this process is a flame produced by the burning of a proper mixture of acetylene and oxygen gases. During the heating process, a filler metal (welding rod) is melted into a prepared groove of the material to be welded.

Oxygen and acetylene are used to produce the welding flame.

Courtesy Kennedy Valve Mfg. Co.

Fig. 1-3. Heat is applied until the flux forms beads.

The combination of these two gases gives a higher temperature (approximately 6000°F) than any other gas flame that could safely be used. For complete combustion a proportion of 2½ parts oxygen to 1 part acetylene is required. However, only 1 part of

Courtesy Kennedy Valve Mfg. Co.

Fig. 1-4. Brazing alloy is applied to the heated area, where it melts and is drawn into the joint by capillary action.

13

oxygen is required to be supplied through the torch because 1½ parts are supplied by the natural air surrounding the flame.

In lighting the torch prior to welding, the acetylene is always turned on first and ignited. The torch will burn with an orange flame and give off black smoke. The flame is now receiving only the 1½ parts of oxygen from the air around it, so combustion is incomplete. When the oxygen is turned on at the torch handle, the additional 1 part of oxygen is provided, and the flame will turn blue with a white inner core, indicating maximum heat and good combustion.

## Oxygen

Oxygen is an odorless, tasteless, and colorless gas that will combine with almost every element known. In combination with certain other substances and under certain conditions, it will produce flame and heat. Oxygen in itself will not burn, but is required for burning to occur. Its property of supporting combustion makes it very valuable in the welding process. Combined with acetylene, it produces one of the hottest gas flames available.

WARNING: *Oxygen under pressure should* NEVER *be permitted to come in contact with grease or oil. The mixture is very explosive and could result in serious injury. That is why the warning* USE NO OIL *is displayed on the regulators for oxygen cylinders.*

Oxygen also results in a rapid rate of oxidation when in contact with hot steel. Considering this factor, no attempt should ever be made to heat a pipe or container holding compressed oxygen.

## Acetylene

Acetylene, which has a characteristic odor and is colorless, is composed of carbon and hydrogen atoms in equal proportions. It is a combustible gas and can form explosive mixtures with oxygen or air.

When calcium carbide is brought into contact with water, a reaction takes place, resulting in the liberation of acetylene gas. Manufacturers state that, because of one of its inherent properties, acetylene gas should never be used at a pressure in excess of 15 psi.

This maximum pressure is sufficient for all cutting and welding operations.

## Welding Rods

An important part in the quality of the finished weld is the welding rod, which is melted into the welded joint. Welding rods of good quality are designed to permit free-flowing metal that will fuse easily with the metal being welded. In this way, good sound welds of correct composition are produced. There are a variety of welding rods available. They are made of cast iron, carbon steel, bronze, aluminum, and stainless steel, as well as other metals.

## Welding Flux

Welding and brazing fluxes are used for the following general purposes:

1. To prevent the molten metal from absorbing and reacting with atmospheric gases.
2. To protect and clean the surfaces of the base metal while the welding is taking place.
3. To purify the molten metal during the welding by removing oxides that are formed.

Flux compositions vary according to the purpose for which they are to be used. Compositions usually are in paste or powder form, and they should be used as recommended by the manufacturer's instructions on the container.

## SETUP AND OPERATION OF EQUIPMENT

Always fasten the cylinders to be used in an upright position so they cannot be knocked or pulled over. If cylinders are not on a suitable cylinder cart, they should be securely fastened, with chain or equivalent, to a workbench, wall, or post. Acetylene cylinders should never be stored or used in other than a vertical position.

"Crack" the cylinder valves. Stand so that the gas leaving the cylinder outlet will not be directed onto your face or clothing. Open the valve quickly about ¼ of a turn, then close it imme-

15

diately. This will clear the valve outlet opening of accumulated dust or dirt that might, if not blown out, mar the seat of the regulator nipple or be carried into the regulator.

PRECAUTION: *Never "crack" a fuel gas cylinder valve near other welding or cutting work in progress or near sparks, flame, or other possible source of ignition. The correct way to crack a cylinder valve is shown in Fig. 1-5.*

Courtesy Union Carbide Co.

Fig. 1-5. "Cracking" a cylinder valve.

Connect the oxygen regulator to the oxygen cylinder and the acetylene regulator to the acetylene cylinder. If the acetylene regulator and the acetylene cylinder have different threads, it will be necessary to use an adapter between the regulator and the cylinder. Two different acetylene cylinder connections are used in the U.S., the CGA 510 connection has left-hand threads, internal on the cylinder outlet; the CGA 300 connection has right-hand threads, external on the cylinder outlet.

Tighten both regulator connections firmly with a wrench, as shown in Fig. 1-6. Wrenches with two or more fixed openings, designed specifically for use with gas welding and cutting apparatus, are available from all apparatus suppliers.

PRECAUTION: *Should it ever be necessary to retighten the regulator union nut after the outfit has been set up and the cylinder valves opened, be sure to close the cylinder valve before tightening the nut.*

**Fig. 1-6. Tightening regular connection with a wrench.**

Rotate the pressure adjusting screw of each regulator to the left (counterclockwise) until it turns freely. The valve in the regulator must be closed before cylinder pressure is applied to the regulator. This is shown in Fig. 1-7.

Open each cylinder valve SLOWLY. Stand where you can see the cylinder pressure gauge, hand on the regulator, but do *not* stand directly in front of the regulator gauge faces. It is especially important that the oxygen cylinder valve be opened only very slightly at first and that you wait until the cylinder pressure gauge has stopped moving before opening the valve fully. The acetylene cylinder valve should furnish sufficient operating pressure at approximately ¾ of a turn open. The acetylene cylinder valve should never be opened more than 1½ turns. When opening the cylinder valves, stand to one side as shown in Fig. 1-5.

17

Courtesy Union Carbide Co.

**Fig. 1-7. The valve in the regulator must be closed before cylinder pressure is applied to the regulator.**

If the acetylene cylinder valve requires a wrench to open it, always leave the wrench in place on the valve while the valve is open. This allows.the valve to be closed quickly if an emergency arises. This wrench is shown, on the valve, in Fig. 1-8.

## Connecting Gas Supplies to the Torch

Always use hose and hose connections made specifically for gas welding and cutting purposes. Oxygen hose has a green cover; acetylene hose has a red cover. Never interchange oxygen and acetylene hose. Do not use acetylene hose with propane fuel unless you know that it is acceptable for use with propane. (Hose with natural rubber liner is satisfactory for acetylene service, but not for propane service.)

Make up all connections dry; do not use pipe-fitting compounds, thread lubricants, oil, or grease. All connections are designed with metal-to-metal seals. they do not require lubricants or sealants. They must always be made up wrench tight, as shown in Fig. 1-9, not hand tight.

Never force connections that do not fit. If you cannot run the

Courtesy Union Carbide Co.

Fig. 1-8. If the acetylene cylinder valve requires a wrench to open it, always leave the wrench in place while the valve is open.

Courtesy Union Carbide Co.

Fig. 1-9. Tightening a hose connection with a wrench.

19

threads together by hand with ease, either the threads are damaged or you are trying to put together parts that were not made to go together.

When the oxygen and acetylene hoses have been connected to the torch, connect the other ends of the hoses to their respective regulators.

Test all connections for leaks. One quick way to do this is to turn the oxygen and acetylene regulator screws in (clockwise) to obtain an oxygen pressure of approximately 25 psi and an acetylene pressure of 7 to 10 psi with the torch valves closed. Close the cylinder valves and observe the gauges. Any leakage between the cylinder valves and the torch will be signaled by a drop in the gauge pressures. A soap-suds solution can be applied to all connections between the cylinder valves and the torch; a leakage at any of these points will be shown by bubbles arising from the soap-suds solution. If no leakage is shown using the soap-suds solution, the leakage may be internal in the regulator or through the throttling valves on the torch. A leak through the valves in the torch may be found by dipping the torch tip in water. If bubbling occurs, open and close the torch valves rapidly to free and blow out any foreign matter that may have lodged in the needle valve openings. If the leakage is still apparent after these measures have been taken, the leaking regulator or torch should be sent to the supplier for repair.

## Lighting the Flame

When lighting the flame, always follow the manufacturer's instruction for the specific torch being used. In general, the procedure for lighting a torch is shown in Fig. 1-10. Note the following:

1. Open the acetylene valve on the torch about ½ turn.
2. *Immediately* light the flame with a friction lighter. NEVER USE A MATCH.
3. Reduce the acetylene flow, by throttling the torch acetylene valve, until the flame just starts to produce black smoke around its edges; then increase acetylene flow just enough to get rid of the black smoke.
4. Open the torch oxygen valve slowly until the desired flame is obtained.

**Fig. 1-10.** Lighting a torch with a friction lighter.

## Flame Adjustment

If oxygen and acetylene working pressures have been set in accordance with the manufacturer's recommendations for the size of the welding tip or cutting nozzle in use, the lighting procedure outlined above will usually produce a neutral flame that has about the right characteristics for most welding and cutting purposes. However, if the flame, when adjusted to neutral, burns away from the tip (if there is a gap between the tip and the flame), reduce the flow of oxygen slightly, using the torch oxygen valve, and then reduce the acetylene flow until the flame is again neutral. If this procedure must be repeated two or three times to eliminate the gap between tip and flame, the acetylene working pressure is probably too high and should be reduced. If the neutral flame produced by following the recommended lighting procedure is too "soft," open the acetylene throttling valve slightly and readjust the flame to neutral by opening the oxygen throttling valve slightly.

21

## Backfire and Flashback

If the welding tip or cutting nozzle should accidentally contact the work, or if the flame should be interrupted momentarily by a droplet of slag, the torch flame may go out with a loud pop. A torch that has backfired may be relighted at once; it may even relight itself if the tip is directed into molten or hot metal. If a torch backfires repeatedly without coming into contact with the work, the cause may be improper operating pressures, a loose tip or nozzle in the torch, dirt on the cutting nozzle seats, or use of too large a welding tip. When an oversized tip is used and the flame is throttled back to attempt to compensate for the oversized tip, backfiring often occurs.

If the flame goes out and burns back within the torch, this usually produces a hissing or squealing noise. If this should happen, shut the torch off immediately. This is called a flashback and indicates either something is wrong with the torch or with the operation of the torch. After a flashback, *always* allow a torch to cool before attempting to relight it. *Always* check the operating pressures before attempting to relight it. Allow the oxygen, *not* acetylene, to flow through the torch before relighting; the oxygen flow will help clear out any soot accumulation. When these procedures have been followed, relight the torch, and if the flame appears normal, proceed with the work. If there is a second flashback, remove the torch from service and return it to the supplier or repair station to be checked and repaired.

## Stopping Work

When stopping work, close the torch fuel valve first, then the oxygen valve. Closing the acetylene valve first reduces the chance of allowing unburned fuel gas to escape and be ignited accidentally. Release all pressure from the torch, hose, and regulators. First close both cylinder valves, then open both torch valves. Finally, release the pressure adjusting screws on both regulators (turning them counterclosewise until they are free) and close both torch valves. Before disconnecting a regulator from a cylinder, always release all pressure from the regulator.

# THE OXYACETYLENE FLAME

The tool used in oxyacetylene welding is not the torch, it is the *flame*. The only purpose of the torch is to provide the gas mixture that will produce the right type of flame for the work being done. There are three types of oxyacetylene flames:

1. A neutral flame.
2. A flame containing excess oxygen (oxidizing).
3. A flame containing excess acetylene (carburizing).

The neutral flame, shown in Fig. 1-11, is produced when the ratio of oxygen to acetylene, in the mixture leaving the torch, is almost exactly one to one. A neutral flame usually has no effect on the metal being welded. It will neither oxidize the weld metal nor will it cause an increase in the carbon content of the weld metal.

The oxidizing flame, shown in Fig. 1-12, results from burning a mixture that contains more oxygen than required for a neutral flame. An oxidizing flame oxidizes or "burns" some of the metal being welded.

The carburizing flame, shown in Fig. 1-13, is created when the proportion of acetylene in the mixture is higher than that required to produce a neutral flame. When a carburizing flame is used to weld steel, it will cause an increase in the carbon content of the metal being welded.

A neutral flame is the proper flame for most types of welding. It is very hard to distinguish the difference between a true neutral flame and a *slightly* oxidizing flame. The difference between a neutral flame and a flame with a slight excess of acetylene is very apparent; the flame with an excess of acetylene will have a definite "feather." When lighting the torch, the flame should be started with an excess of acetylene, then increase the oxygen flow until the acetylene feather just disappears. If the flame is too large for the work being done, reduce the oxygen flow first, to produce a feather, then cut back on the acetylene flow until the feather just disappears. Instructions for some other welds may call for an oxidizing flame; other welds may call for a carburizing flame. The implication here is that an excess of one type of gas will not cause any trouble with the weld but that an excess of another gas may cause trouble.

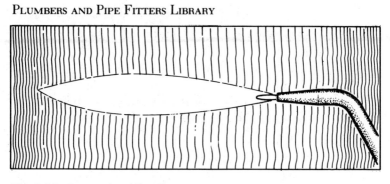

Fig. 1-11. A natural welding flame.

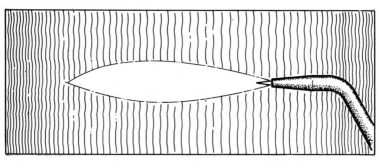

Fig. 1-12. An oxidizing welding flame.

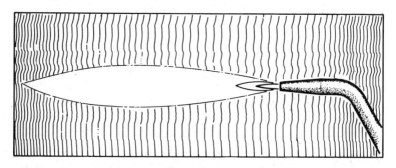

Fig. 1-13. A carburizing welding flame.

24

# WELDING TORCHES

A welding torch consists of a handle containing throttling valves for the oxygen and acetylene and a *tip*. Fig. 1-14 shows one type of torch that is widely used by plumbers and pipe fitters in their work. Various sizes of tips are available for welding a variety of pipe sizes.

## THE CUTTING TORCH

The cutting torch shown in Fig. 1-15 is made to fit the handle of the welding torch shown in Fig. 1-14. The throttling valves on the handle control the oxygen and acetylene used in the cutting torch; in addition, there is a throttling valve to control only the oxygen used when cutting. The preheat flame of the torch is adjusted to a neutral flame by using the same procedure as outlined for the welding torch; the throttling valve for the lever-operated cutting valve is closed while adjusting the preheat flame. When the preheat flame is adjusted, the *throttling valve* on the cutting torch is opened and the lever-operated *cutting valve* is depressed or opened until the proper oxygen flow for cutting is obtained. Setting the cutting torch is not as complicated as it might seem. The skilled welder can adjust the cutting torch in a matter of seconds.

## REGULATORS

Regulators for oxygen and acetylene are made in two types: single-stage regulators and two-stage regulators. The two-stage regulator is preferred by most welders; it delivers a constant flow regardless of cylinder pressure, will stand up better in constant service, and requires less maintenance than single-stage regulators. When a regulator has been connected to an oxygen or acetylene cylinder and tightened, the cylinder valves should be opened as shown in Fig 1-7. They should always be opened *slowly* to prevent damage to the regulator and the regulator gauges. The valve on the oxygen cylinder should be fully opened when in use; the valve seats when fully opened to prevent leakage around the stem. The acetylene valve should only be opened slightly, ¾ to 1½ turns, so that it may be closed quickly in the event of an accident. The adjusting valves on the regulators, both oxygen and

Courtesy Union Carbide Co.

**Fig. 1-14. One type of torch used by plumbers and pipe fitters.**

Courtesy Union Carbide Corp.—Linde Division

**Fig. 1-15. A cutting torch used with the welding torch shown in Fig. 1-14.**

acetylene, should be closed (turned counterclockwise until they are free) before turning the cylinder valves on. Fig. 1-7 shows how the pressure adjusting screw or valve should be released (closed) before turning the cylinder valve on. A two-stage regulator is shown in Fig. 1-16.

Courtesy Union Carbide Corp.—Linde Division

**Fig. 1-16. A two-stage regulator with gauges.**

## PIPE WELDING

Welding pipe is used extensively in heating and air-conditioning systems. Certain techniques and methods of welding and aligning pipe for welding will help the novice welder to become a skilled pipe welder.

The ends of pipe to be joined by welding should be beveled correctly. Weld-type pipe and fittings are made with ends beveled to the correct angle. When pipe is cut on the job, the welder must use a cutting torch to bevel the ends of cut pieces. Fig. 1-17 shows the correct method of making a beveled cut. The beveled cut should be at an angle of approximately 35°; the two cuts placed in position for welding should produce a total angle of approximately 70° to be filled when welding is completed.

Correct alignment of two pieces of pipe to be welded can be ensured by placing the two pieces in a section of angle iron while the tack welds are being made, as shown in Fig. 1-18. When long

27

Fig. 1-17. Correct position for making a beveled cut on pipe.

Fig. 1-18. Using angle iron to align pipe for welding.

sections of pipe are to be joined by welding, the use of a welding clamp, shown in Fig. 1-19, is advised. This clamp will secure the two pieces in position and alignment while the tack welds are made. After the first tack weld is made, the pipe should be rotated 180°, and the second tack weld should be made opposite the first one; the pipe is rotated another 90° to make the third tack weld, and yet another 180° for the fourth tack weld. If the tack welds are made in this sequence, the possibility of misalignment due to contraction of the welds will be minimized.

**Fig. 1-19. A pipe-alignment clamp.**

Fig. 1-20 shows the correct method of starting the finished weld. The weld should be started about 1" away from a tack weld. When working at a bench, a pipe roller is very useful since it permits the weld to be made from one position. A very good pipe roller can be made from a discarded pair of roller skates. As shown in Fig. 1-20, completing this weld will involve welding in the flat, vertical, and overhead positions. Although the use of a pipe roller will make some welds easier and quicker to make, the skilled pipe welder must be able to work in all positions. The exact starting point of the weld and the points at which the welder should change his position relative to the weld are matters of personal preference.

29

Courtesy Union Carbide Corp.—Linde Division

**Fig. 1-20. Starting the finished weld.**

Fig. 1-21 shows the root pass of a 4″ pipe weld well started. If the pipe wall is more than ⅛ inch thick, two passes are recommended. For thinner-wall pipe, only one pass is needed. Note that in both Fig. 1-20 and Fig. 1-21 the torch flame is pointed almost, but not

Courtesy Union Carbide Corp.—Linde Division

**Fig. 1-21. Continuing the weld—welding toward the top (forehand method).**

30

quite, toward the center line of the pipe, while the filler rod is held approximately tangent to the pipe surface. This is the normal angle for pipe welding, regardless of the welding position, except when adjustment of flame angle is required for puddle control in the horizontal and overhead positions. Unless the pipe ends have been prepared perfectly, there will be variations in the width of the joint gap at the root. This may require some careful bridging, by concentrating first on one side of the joint, then on the other. One of the big advantages in oxyacetylene welding is that heat input and filler metal input can be controlled separately. Gaps that can give the arc welder trouble are no problem at all to the experienced oxyacetylene pipe welder.

Proper rod movement is the key to putting the finishing pass on a pipe weld. The rod must rub in the bottom of the puddle and bump against the edges of the joint. The flame must linger a bit at each end of each arc and be concentrated on the puddle—never on the filler rod. Movement of the flame and rod must be in opposite directions. As shown in Fig. 1-22, the welder is working in the

Courtesy Union Carbide Corp.—Linde Division

**Fig. 1-22. Welding in the vertical position.**

vertical position. The flame is pointed somewhat away from the center line of the pipe in order to hold the puddle against the force of gravity.

Fig. 1-23 shows the welder using the backhand technique. In pipe welding, when the pipe cannot be rotated, there are many occasions when the ability to switch easily from forehand to backhand can save time for the welder. With the backhand technique it is quite easy to make a vertical weld from top to bottom, rather than from bottom to top. The flame is still used to "support" the puddle, but in this case its aim is to keep the puddle from running ahead on to metal that has not yet reached melting temperature. When using the backhand technique, the relative movements of flame and rod are different from those used in forehand work. The novice welder is advised to practice the backhand technique on flat and vertical pieces of plate before employing the technique on pipe.

It is often necessary to make a weld in a section of vertical piping. Fig. 1-24 shows the root pass being made on this type joint. In making the root weld, keep the puddle as small as possible and angle the flame slightly toward the upper side of the joint. The rod should be held in what is considered the normal position for all pipe work, approximately tangent to the circumference of the pipe. When the finish pass is made, the puddle must be slanted slightly with the lower end leading the upper end, and the filler rod must be manipulated to push molten metal up against the top edge of the joint.

Fig. 1-24 shows the advantages in being able to weld using both the forehand and backhand techniques. If the weld is made using only the forehand method, the weld must be started at the bottom of the joint and worked up one side; then the welder must change position, start at the bottom again, and work up the other side. If the backhand method has been mastered, the welder can start on one side, work down to the bottom using the backhand technique, switch to the forehand method, and work up the other side. The ability to switch from forehand to backhand, and vice versa, can be a big timesaver when welding fixed joints in horizontal pipe runs. As shown in Fig. 1-25, an adjustable support using a sliding clamp to which a C clamp has been welded is excellent for use when practicing overhead welding. Becoming a good pipe welder

Courtesy Union Carbide Corp.—Linde Division

Fig. 1-23. Finishing the weld—using the backhand method.

Courtesy Union Carbide Corp.—Linde Division

Fig. 1-24. Making a weld in a section of vertical piping.

33

**Fig. 1-25.** Welding fixed joints in horizontal pipe runs. The ability to make both forehand and backhand welds is very useful in this type of weld.

is largely a matter of having good instruction and good tools—and constant practice until the art of welding is mastered.

## GENERAL PRECAUTIONS

CAUTION! IMPORTANT! USE NO OIL! Oil, grease, coal dust, and some other organic materials are easily ignited and burn violently in the presence of high oxygen concentrations. These materials should never come in contact with oxygen or oxygen-fuel gas equipment. Lubrication of any type is *never* required with oxygen-fuel gas apparatus.

*Always call oxygen by its proper name! Do not call it "air."* Never use it as a substitute for compressed air. A serious accident may result if oxygen is used as a substitute for compressed air. Oxygen must never be used in pneumatic tools, in oil preheating burners, to start internal combustion engines, to blow out pipelines, to "dust" clothing or work, for pressure testing, or for ventilation.

*Never use acetylene at pressures above 15 psi.* The use of acetylene at pressures in excess of 15 psi is extremely hazardous.

*Never use torches, regulators, or other equipment that is in need of repair.* A "creeping" regulator, leaking torch valves, or leaking hose connections usually require repair by a qualified repair station.

*Never connect an oxygen regulator that is not equipped with an inlet filter to a cylinder.*

*Always use the operating pressures recommended by the manufacturer for the welding head or cutting nozzle in your torch.*

*Always wear goggles when working with a lighted torch.* The choice of shades will be up to the welder and will generally be in the range of shades 4, 5, and 6.

*Do not use matches for lighting a torch.* Always use a friction lighter or some other source of ignition that will keep your hands away from the torch tip or nozzle.

*Wear suitable clothing when using a torch.* Fire-resistant gauntlets, pants free from oil or grease, pants without cuffs, and high-top shoes should all be worn while operating a torch.

*Before starting to weld or cut, check the area to be sure that sparks or slag will not start a fire or damage walls, ceilings, floors, furniture, or equipment.*

*Never do welding or cutting without adequate ventilation.* The four paragraphs that follow are reprinted from *The Oxy-Acetylene Handbook*, published by the Union Carbide Corporation-Linde Division.

1. **When it is necessary to work in a confined space, make certain that the space is adequately ventilated.** *If there are no overhead openings to provide natural ventilation, use an exhaust fan or blower. When screening an area, try to keep the bottom of the screens at least two feet above floor level. Do not take cylinders into a confined space. Test all equipment, including hose, for leaks before taking it into a confined space, and bring it out with you when your work is interrupted for any reason, even for a short time.*

2. **Never flow oxygen into a confined space in order to ventilate it, or to "clear the air."** *Remember that oxygen supports and accelerates combustion, and will cause oil, wood, and many fabrics to burn with great intensity. Clothing saturated with oxygen, or with oxygen enriched air, may burst into flame when touched by a spark.*

3. **Do not weld brass, bronze, or galvanized steel except in a well-ventilated location.** *You must protect yourself against breathing the zinc oxide vapors usually generated when these materials are heated to welding temperatures.*

4. **When welding or cutting metals containing or coated with lead, cadmium, beryllium, or mercury, always wear a suitable air-line mask.** *No operator should be considered immune from the effects of these fumes. A straight filter-type mask is inadequate; nothing short of an air supplied respirator is suitable. There must be sure protection against breathing fumes that occur when lead, mercury, beryllium, or their compounds are heated.*

*Use particular caution when welding or cutting in dusty or gassy locations.* Dusty and gassy atmospheres in some mines and plants call for extra precautions to avoid explosions or fires from sparks, matches, or open flames of any type. Welding or cutting in any such "suspicious" location should be done only when proper precautions are taken, and only after a responsible official has inspected the situation and given approval for the work.

When welding or cutting is done underground, leak testing of equipment setups should be performed carefully and frequently. Great care should be taken to protect hoses and cylinders from damage, and to protect timbers and other combustible materials from sparks and hot slag.

*Never do any welding or cutting on containers that have held flammable or toxic substances until the containers have been thoroughly cleaned and safeguarded.* Complete, detailed procedures are given in American Welding Society booklet A6.0-65, *Safe Practices for Welding and Cutting Containers That Have Held Combustibles.* Some of the key points in this booklet are:

1. Always start with the assumption that any used drum, barrel, or container may contain flammable (explosive) or toxic (poisonous) residue.

2. No work should be begun until the container has been cleaned and tested sufficiently to assure that no flammable or toxic solids, liquids, or vapors are present.
3. Bear in mind that some nonpoisonous substances can give off toxic vapors when heated.
4. Steam is usually an effective means of removing volatile or readily volatilizable materials from a container. Washing with a strong solution of caustic soda or other alkaline cleaner will remove heavier oils that steam may not volatilize.
5. Whenever possible, the container should be filled with water to within a few inches of the working area before attempting welding, cutting, or intense heating. Care should be taken to provide a vent opening for the release of heated air or steam.
6. When it is impractical to fill the container with water, an inert gas such as nitrogen or carbon dioxide may be used to purge the container of oxygen and flammable vapors. Maintain the inert gas during the entire welding or cutting operation by continuing to pass the gas into the container.

**Other Safe Practices to Bear in Mind**

*Make sure that jacketed or hollow parts are sufficiently vented before heating, welding, or cutting.* Air or any other gas or liquid confined inside a hollow part will expand greatly when heated. The pressure created may be enough to cause violent rupture of the part. A metal part that is suspiciously light is hollow inside and should be drilled to vent it before heat is applied. After work is complete, the vent hole can be tapped and plugged if necessary.

*Bushings in a casting should either be removed or securely fastened in place before heating the casting.* Bronze bushings expand more than cast iron when heated to the same temperature. If a bushing is left in place, the casting may be damaged, or expansion may cause the bushing to fly out, creating a definite hazard. If a bushing cannot be removed, it should be securely fastened in place. Bolting large washers or pieces of plate over the ends of the bushing is the suggested method.

*Don't drop stub ends of welding rods on the floor. Put them in a suitable container.* Aside from the fire hazard that may be created by carelessly dropping a hot stub end, a serious fall may result from stepping on one. A small container partly filled with water and within easy reach is a good place to dispose of these short ends. Protect cylinders, hose, legs, and feet when cutting. Do not cut material in a position that will permit sparks, hot metal, or the severed section to fall on or against a gas cylinder or the gas hoses or your legs or feet. A suitable metal screen in front of your legs will provide protection against sparks and hot metal.

## Preventing Fires

*Where welding or cutting has to be done near materials that will burn, take special care to make certain that flame, sparks, hot slag, or hot metal do not reach combustible material and thus start a fire.* This is particularly important in the case of cutting operations. Cutting produces more sparks and hot slag than welding, and locations where portable cutting equipment is used must be thoroughly safeguarded against fire.

*Never use cutting or welding torches where sparks or open flame would be a hazard.* Flames are a hazard in rooms containing flammable gas, vapors, liquids, dust, or any material that easily catches fire. It is not safe to use cutting or welding equipment near rooms containing flammable materials unless there is absolutely no chance of sparks passing through cracks or holes in walls or floors, through open or broken windows, or through open doorways. There is always the possibility that flammable vapors may escape from such rooms through doors or other openings, making it doubly necessary to keep sparks and flames away.

*Before you cut or weld in a new location for the first time, always check with the nearest person in authority.* He may know of some serious fire hazard that might otherwise be overlooked.

*If the work can be moved, take it to a location where there will be no possibility of setting fires.* This must always be done when the metal to be welded or cut is in a place where open flames are barred. This practice may also be sensible in other locations even if open flames are permitted.

*If the work cannot be moved, materials that burn easily should, if possible, be taken a safe distance away.* For cutting operations, this distance may be 30' to 40' or more. Floors should be swept clean before the torch is lighted.

*If flammable materials cannot be moved, use sheet-metal guards, asbestos paper or curtains, or similar protection to keep sparks close in to the work you are doing.* Suitable protection to keep back sparks should always be used when it is not possible to move materials that will burn to a safe distance from the cutting or welding work. This also applies if sparks might lodge in wooden parts of the building or drop through holes or cracks to the floor below. Make sure that the guards are large enough and tight enough so that they do not permit sparks to roll underneath or slide through openings. Curtains should be weighted down against the floor or ground. For weights, use such things as angle iron, pipe, bricks, or sand. Use only fire-resistant guards. Do not use tarpaulins for shielding sparks since they may catch fire. Have someone stand by to watch the sparks so that he can give warning if they begin to get beyond the protective guards. It is not reasonable to expect whoever is doing the welding or cutting to watch the sparks because his attention is on the work. In addition, the sparks cannot always be seen easily through goggles.

*If welding or cutting over a wooden floor, sweep it clean and wet it down before starting work.* Provide a bucket or pan containing water or sand to catch the dripping slag from any cutting that is done.

*Before starting to cut off a piece of steel or iron, make sure it will not drop where there is any possibility of starting a fire.* This is especially important when working in a high place where sparks or slag might be less likely to cause a fire down below that would a small piece of red hot steel. Pieces can be kept from falling by welding a rod or bar to the piece, and having a helper hold the rod or bar while the cut is made. Or the same thing might be accomplished by tying a chain or other suitable support to the piece.

*When welding very close to wooden construction, protect it from direct heat.* Wooden beams, partitions, flooring, or scaffolding should always be protected from the direct heat of the flame by sheet-metal or other guards. Protective guards should also be used to confine the sparks.

*Use the correct oxygen pressure when cutting.* An oxygen pressure greater than necessary will only cause extra sparks and increase the slag flow, to say nothing of increasing the oxygen expense.

*Store extra cylinders away from important areas.* Keep only enough cylinders near the work to ensure an adequate supply of gases for the job at hand.

*Be ready to put out any fire promptly with fire extinguishers, pails of water, water hose, or sand.* When torches are to be used near wooden construction or material that will burn, take every precaution to prevent fires, but always be prepared to put out fires that may start despite all precautions. In hazardous locations, have a helper, or one or more extra men if necessary, on hand to watch for and be ready to extinguish a fire.

*If there is a sprinkler system in the building, maintain this protection without fail while cutting or welding is being done.* It is of special importance to make sure that sprinklers are in working order during extensive repairs or building changes. If the sprinkler system must be shut down for a time, have this done when welding or cutting work is not in progress.

*If there is a possibility that a smoldering fire may have been started, keep a man at the scene of the work for at least a half hour after the job is through.* Have him look carefully for smoke or fire before leaving. This is especially important when cutting torches have been used in locations where sparks may have started smoldering fires in wooden structures or in other slow burning materials. *Never forget* that heavy cutting sparks sometimes fly 25′ to 30′ or more and hold their heat for several seconds after landing.

*When welding or cutting on bridges, structures, or at other outdoor sites, take care to avoid setting fire to grass or brush.* Brush should be cleaned out or cautiously burned under or about structures before start of the work. Special care is necessary during a dry spell. A fire extinguisher and water or sand should be available to extinguish any fires started in the course of work. Before leaving the site, examine the premises thoroughly to be sure that sparks have not started smoldering fires.

## ARC WELDING

Many different types and sizes of arc-welding machines are available, but all fall into two basic categories—the conventional, or *constant-current,* welding machine with the drooping volt-ampere curve, and the *constant-voltage* or modified constant-voltage machine with the fairly flat characteristic curve. The conventional machine can be used for manual welding and, under some conditions, for automatic welding. The constant-voltage machine is used only for semiautomatic and automatic welding. Another classification of welding machines is based on whether they are rotating types or not. The rotating or generator type is the most common. This machine gives a wider range of welding-current adjustment and a wider range of electrodes. In shop work, the generator is powered by an electric motor; in the field, by any internal-combustion engine. This machine, because it has its own power source, is a good practical machine for emergency work, and when away from available power lines.

Rotating generators are available to provide either DC or AC. In some cases, particularly when driven by a gasoline or diesel engine, the generator is designed to produce AC and thus can be used to provide power for lights and tools as well as power for welding. The AC is rectified in some machines to provide DC.

## Conventional Single-Operator Welding Machines

This type of machine, called a *constant-current* welder, produces a volt-ampere curve like the one in Fig. 1-26. This curve is obtained by loading the machine with variable resistance and plotting the terminal voltage for each ampere output. A brief study of the curve will reveal that a machine of this type produces maximum output voltage with no load, and as the load increases, the output voltage decreases. For normal welding conditions, the output voltage is usually from 20 to 40 volts. The open-circuit voltage is between 60 and 100 volts. This high no-load voltage provides for easier arc starting with all types of electrodes. The open-circuit, or no-load, voltage is controlled by the fine adjustment on a dual-control welding machine.

The welding current is adjusted by both the range switch for coarse adjustment and a fine adjustment. On this type of machine, the actual arc voltage is controlled by the welding operator and has a direct relationship to the arc length. As the arc length is increased, the arc voltage increases—decreasing the arc length decreases the arc voltage.

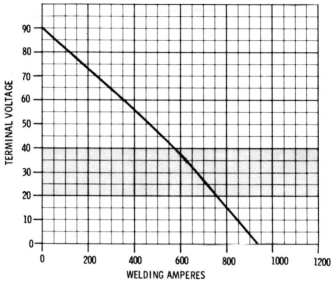

**Fig. 1-26. Volt-ampere curve of a constant-current welder.**

A study of the curve will reveal that with a long arc (higher arc voltage), the current output will decrease a limited amount, whereas with a shorter arc (low arc voltage), the current output will increase a limited amount. Thus, the operator can vary the current output or heat of the machine by lengthening or shortening the arc.

On dual-control machines, the slop of the output curve can be varied. In this way, a soft or harsh arc can be obtained. With a flatter curve, the same change in arc voltage will produce a greater change in current output. This produces a digging arc preferred for pipe welding. With a steeper curve, the same change in arc voltage will produce less change in the output current. This is a soft, or quiet, arc useful for sheet-metal welding. In other words, a conventional or constant-current welder with dual controls allows the most flexibility for the welding operator. These machines can be driven by an electric motor or internal-combustion engine.

A typical electric-drive arc welder is shown in Fig. 1-27. Machines of this type have multirange dual controls, which solve the problem of getting just the right heat for each and every job.

**Fig. 1-27. A portable constant-current arc welder.**

The large handwheel on the front of the control panel enables the operator to choose any one of ten coarse ranges. The volt-amp adjuster in the center of the wheel permits up to 100 fine adjustments for each coarse range, providing 1000 possible welding combinations. This welder also has a remote-control device that permits the operator to make adjustments directly at the welding site.

## Constant-Voltage Welding Machines

A constant-voltage welding machine provides a practically constant voltage to the arc, regardless of the amount of arc current. This type of welder is shown in Fig. 1-28. The characteristic curve of this machine is shown by the volt-ampere curve in Fig. 1-29.

A constant-voltage welder can be used only for automatic or semiautomatic welding, with one exception. With the proper voltage-dropping ballast resistors, this type of machine can be used to supply power simultaneously to a number of welding operators. However, the constant-voltage power supply was designed specifically for automatic welding. It is a known fact that the burn-off rate of a specific size and type of electrode wire for a given arc voltage is proportional to the welding current. In other words, as the welding current increases, the amount of wire burned off will increase proportionally. This is shown graphically on the curve in Fig. 1-30. Thus it can be seen that if a wire is fed into an arc at a specific rate, it will automatically require, or draw, a proportionate amount of current from a constant-voltage power source.

Courtesy Hobart Brothers Co.

Fig. 1-28. A constant-voltage arc-welder.

A constant-voltage welder provides the amount of current required from it by the load imposed on it. In this way, a very basic automatic welding control can be employed. The wire is fed into the arc by means of a constant-feed motor. This motor can be adjusted to increase or decrease the rate of wire feed. Complicated circuitry is thus eliminated, and the system is inherently

Courtesy Hobart Brothers Co.

Fig. 1-29. Volt-ampere curve of a constant-voltage welder.

Courtesy Hobart Brothers Co.

Fig. 1-30. Burn-off rate curve of an automatic welder.

self-regulating. If the electrode wire is fed in faster, the current will increase. If it is fed in slower, the current will automatically decrease. The current output of the machine is thus set by the speed of the wire-feed motor.

The voltage of the generator is regulated by an output rheostat on the generator proper. Thus, only two controls maintain the proper welding current and arc voltage in a constant-voltage system. It is almost impossible to have stubbing or burn-back, which is common with the constant-current welding machine.

The characteristic curves of constant-voltage machines have a slight inherent droop. This droop can be increased (the slope made steeper) by various methods. Most machines have taps for varying the slope of the characteristic curve. It is important to select the slope most appropriate for the work being welded.

## PIPE WELDS

### (Courtesy Crane Co.)

There are numerous types of welds used for various purposes and different kinds of pipe connections. Those in general use are as follows.

### Butt Welds

Two types of butt welds are generally used to join pipe sections. The one in Fig. 1-31A is for pipe with a wall thickness of ¾ " or less. For pipe thickness greater than ¾ ", the butt weld shown in Fig. 1-31B is recommended. Figs. 1-31C and D show welds utilizing a machined backing ring, and with an inert arc used on the first pass.

### Welding Couplings and Pads

Smaller sizes of screwed and socket-weld couplings and connections are generally prepared in one of two ways—welding couplings (Fig. 1-32A and B) and welding pads (Fig. 1-32C).

### Welded Nozzles

In the fabrication of pipe with welded nozzles, two types are utilized—the *standard type*, which is not reinforced; and the *rein-*

46

(A) For metal thickness of ¾" or less.

(B) For metal thickness greater than ¾".

(C) With machined backing ring for any metal thickness.

(D) For all metal thicknesses with inert arc first pass.

Courtesy Crane Co.

Fig. 1-31. Typical butt welds used to join pipe.

(A) Threaded coupling.

(B) Coupling bored for socket welding.

(C) Threaded pad.

Courtesy Crane Co.

Fig. 1-32. Welded couplings and pads.

*forced type,* which provides reinforcement of the nozzle weld. Fig. 1-33A shows a 90° standard type, while Fig. 1-33B shows a 90° reinforced nozzle. Angular types with standard and reinforced welds are shown in Figs. 1-33C and D.

Reinforcing of welded nozzles is done as required by the customer and in accordance with ASMF Boiler and Pressure Vessel Code or the ASA Pressure Piping Code when compliance with these codes is mandatory. The ring reinforcement is furnished in the absence of definite specifications for fabricated piping with reinforced welding nozzles.

## Miter Weld

A miter weld (Fig. 1-34) may be used for fabricating pipe in larger sizes or for unusual conditions. Where possible, the use of

47

(A) Standard 90° nozzle.

(B) Reinforced 90° nozzle.

(C) Standard angular nozzle.

(D) Reinforced angular nozzle.

Courtesy Crane Co.

**Fig. 1-33. Typical angular and 90° nozzle welds.**

**Fig. 1-34. A miter weld.**

Courtesy Crane Co.

pipe bends or welding elbows is recommended in preference to miter welds. Where space or design limitations require, or when suitable welding elbows or bends are not available, miter welding can be used. Such welds are made up of various combinations of segments, angles, and so on, and must be designated for each condition encountered.

# MICRO-WIRE PIPE WELDING

### (Courtesy Hobart Brothers Co.)

The piping industry, a major user of welding, is always looking for ways to reduce costs. A method developed by Hobart Brothers Company, called the *Micro-wire* welding process, was introduced to meet these demands. It was promptly investigated by the piping industry, and after extensive laboratory work and field experiments, was used in actual production in pipe welding.

Micro-wire welding is a gas-shielded, arc-welding process that uses a small-diameter, consumable electrode wire that is continuously fed into the arc. Metal is transferred through the gas-protected arc column to the work. The process may be either automatic or semiautomatic, the latter method being the more popular. The diagram of the process in Fig. 1-35 shows the electrode wire and the shielding gas being directed into the arc by a lightweight, manually controlled welding gun.

## Advantages

Many distinct advantages of Micro-wire over manual coated-electrode welding are claimed by the manufacturer. Some of these are:

1. Reduce welding time per pipe joint.
2. Elimination of flux and slag, thus reducing the cleaning time per weld joint.

GAS SHIELDING

SMALL-DIAMETER FILLER WIRE

**Fig. 1-35. Micro-wire gas-shielded metal arc.**

**Courtesy Hobart Brothers**

49

3. Fewer passes required per joint.
4. Fewer starts and stops per weld pass.
5. Usable on almost all sizes of pipe and pipe-wall thickness.
6. High-quality, "low-hydrogen" type of weld metal.
7. Smooth weld surface produced inside the pipe.
8. Back-up rings normally not required.
9. Training period for welding operators shorter than for manual coated-electrode welding operators.
10. Weld appearance superior to that produced with manual coated electrodes.
11. No electrode stub loss.
12. Distortion and warpage reduced, especially on thin-wall pipe.
13. Initial or root pass thicker and stronger, allowing faster "move-up."
14. Usable for hot tapping.

There are many other minor advantages claimed for Micro-wire welding, such as smoothness of the exterior surface, ease of painting, and reduced warpage.

## Piping Categories

The pipe industry is divided into four distinct groups. Each has its own specific requirements and is governed by different conditions and specifications. These four groups are:

1. Subcritical piping.
2. Pressure or power piping.
3. Cross-country transmission-line piping.
4. Water-supply piping.

*Subcritical* piping consists mostly of small-diameter pipe applications, such as domestic water supply, sanitary systems, and air conditioning. No code or qualification procedures cover subcritical piping.

*Pressure and power piping* is used for steam generating, refining, and chemical processing. Welding for this type of piping is governed by the ASA Pressure Piping Code, which is based on Section IX of the ASME Pressure Vessel Code. When properly applied, Micro-wire welding satisfies these requirements.

*Cross-country transmission piping* carries fuel gas and/or petroleum products over long distances. Welding of such piping (usually of large diameter) is governed by the American Petroleum Institute Standard 1104 Code. Micro-wire welding qualifies for this type of piping also, when properly applied.

*Water-transmission piping* is not governed by any code except that specified by the owner, which is usually a local government code. This piping is of large diameter, often as large as 96'.

Typical companies that use the Micro-wire process for welding pipe are as follows:

Cross-country pipeline contractors.
Mechanical contractors.
Utility companies.
Heating, ventilating, and air-conditioning contractors.
Refineries.
Chemical plants.
Food processing industries.
Beverage industries.

## Application

Micro-wire welding features versatility and economy. Welding is possible in all positions, the arc is visible to the operator, there is no slag to remove, and the weld surface is smooth. The process is used primarily for welding low- and medium-carbon steels, and low-alloy, high-strength steels of thin to medium gauge. Welding can be performed on extremely thin materials (20-gauge) and will successfully bridge wide gaps. The Micro-wire process has produced highly satisfactory welds in products of the automotive and household appliance industries and in the fabrication of metal furniture, fuel tanks, and pressure vessels. Other applications include electrical equipment, industrial machinery, pipe joining, salvage operations, and highway, bridge, and structural fabrication.

## Equipment

Fig. 1-36 shows the more important components of a Micro-wire welding outfit. Included are the shielding-gas system and controls, the welding machine (power source), the wire-feeding

51

Fig. 1-36. Major components of a Micro-wire welding system.

mechanism and controls, the electrode wire, and the welding gun and cable assembly.

Various shielding gases can be used. Carbon dioxide $CO_2$ gas is the least expensive and the most popular. Mixtures of $CO_2$ and argon and mixtures of argon and oxygen, are also sometimes used. The $CO_2$ gas should be of *welding grade*, meaning that it should have a low moisture content, indicated by the *dew point temperature*.

A dew-point temperature of $-40°F$ or lower is recommended. The gas-flow rate is very important. A pressure-reducing regulator and a flowmeter are required on the gas cylinder. For Micro-wire welding, the flow rate is from 15 to 20 cubic feet per hour as measured and set by the flowmeter. When welding outside, or when air currents disturb the gas shield, a higher gas-flow rate is necessary, sometimes for 30 to 35 cubic feet per hour. To maintain the higher rate, two or more gas cylinders are manifolded together. When high flow rates are used, it is sometimes necessary to heat the gas by means of a gas heater.

A specially designed *welding machine* (power source) is used for Micro-wire welding. It is a constant-voltage type. It can be a DC rectifier or generator, with the generator motor—or can be engine driven. The output welding power of a constant-voltage machine has essentially the same voltage no matter what the welding current may be. The output voltage is regulated by a rheostat on the welding machine. Micro-wire welding uses an arc voltage of from 16 to 24 volts; the open-circuit voltage (while not welding) and the arc voltage (while welding) are read on the welding machine's voltmeter. There is no current control on a constant-voltage machine, and for this reason, it *cannot be* used for manual electrode welding. The welding-current output is determined by the wire feeder.

The specially designed *wire-feeding mechanism* and the constant-voltage welding machine constitute the heart of the Micro-wire welding process. There is a fixed relationship between the rate of electrode wire burn-off and the welding current. In other words, at a given wire-feed rate, the welding machine will produce the current required to maintain the arc. Thus, the electrode wire-feed rate determines the welding current. Wire-feed speed (thus, the welding current) is set by the wire-feed speed control on the wire feeder. The welding current, which can be read on the welding machine's ammeter, ranges from 50 to 300 amperes for Micro-wire welding. Reverse polarity is used with direct current.

The electrode wire is, normally, 0.030″, 0.035″, or 0.045″ in diameter. It is solid and bare, except for a very thin copper coating, or preservative, on the surface to prevent rusting. It contains dioxidizers that help to clean the weld metal and to produce sound, solid welds. The composition of the electrode wire must be matched to the base metal being welded.

The *welding gun and cable assembly* is used to carry the electrode wire, the welding current, and the shielding gas from the wire feeder to the arc area. The operator directs the arc and controls the weld with the welding gun. The distance from the tip

Courtesy Hobart Brothers

Fig. 1-37. A portable Micro-wire welding system driven by an electric motor.

of the gun to the work is called the *stickout*. This stickout distance is controlled by the operator, but the arc length is automatically controlled by the voltage setting on the welding machine. The stickout can be varied somewhat, and is useful in controlling the weld. Too much stickout reduces the gas shield and must be avoided. A stickout (tip-to-work) distance of ¼" to ⅜" is best. The gun nozzle must be cleaned regularly to assure efficient gas shielding. Special antisplatter compounds are used frequently to prevent a spatter buildup on the nozzle.

The equipment required for automatic Micro-wire welding is essentially the same as for the semiautomatic process, except that fixturing is provided and the welding gun is replaced by an automatic welding torch.

Courtesy Hobart Brothers

Fig. 1-38. A Micro-wire welding system powered by a gasoline engine.

Fig. 1-37 shows a typical Micro-wire semiautomatic wire-feed welding machine used where a power source is available. Fig. 1-38 shows a generator welder rated at 350 amperes using a water-cooled gas engine for in-the-field use.

CHAPTER 2

# Planning a
# Heating System

Heating is used in nearly every section of the United States, and in most areas is required for most of the year. In some instances, the same equipment is used for summer cooling as for heating during the cooler seasons. When both heating and cooling are combined in an integrated unit, the functions required for both must be considered when planning for the heating system alone.

The first step in determining the size of a heating system is to accurately determine the heat loss of the building. Improved construction methods and building materials have negated the rule-of-thumb method formerly used to determine heat loss. To avoid oversizing the heating plant, with consequent increase in installation costs, every effort should be made to obtain as accurate a heat-loss determination as possible.

When the outside temperature is lower than the inside temperature, heat is lost through floors, ceilings, walls, doors, and windows. Windows and doors allow some infiltration of air through cracks. Thus heat is lost to the outdoors in two ways—by infiltration around doors and windows, and by transmission, or transfer, through the building materials.

## HEAT LOSS PER SQUARE FOOT

### Step 1

To properly determine the amount of heat lost through a particular type of construction, it is necessary to work with a specific value known as a *coefficient of heat transmission*. This value is designated as the *U factor*, which represents the time rate of heat flow (expressed in Btu/hr) for 1 square foot of surface with a temperature difference of 1°F between the air on one side and the air on the other. "U" factors have been determined for a wide variety of wall, floor, ceiling, window, and door construction. These are available from the American Society of Heating & Ventilating Engineers as well as many manufacturers of heating equipment.

The location of the walls, ceilings, windows and skylights, floors, and doors are listed on a tabulation sheet (Fig. 2-1) along with the type of construction. Walls may be located according to their exposure (North, West, etc.), by the room, or both.

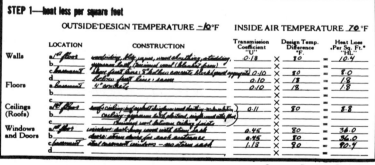

Fig. 2-1. A sample tabulation sheet used to calculate the heat loss per square foot of exposed area.

*NOTE: Heat loss is calculated for all surfaces exposed to the outdoors, to unheated spaces, or to partially heated spaces.* List the "U" factor for each entry on the sheet.

*NOTE: It is common practice in calculating the transmission heat loss of a door to use the "U" factor for a comparable size single-pane glass window. This is because accuracy in selecting the proper "U" factor for the actual door construction varies widely. By using the glass factor, the heat-loss calculation will be on the safe side.*

**Design Temperature.** The heat-loss calculation also depends on the desired inside air temperature. 70°F is usually taken to be adequate, but some building owners may require a higher temperature. The outside temperature depends on geographical location. The coldest outside temperature expected during a normal heating season is known as the *outside design temperature.* It is not necessary to install a heating system capable of meeting the lowest temperature on record because such a temperature may never be reached again, or even approached except in rare instances. In design practice, a temperature is selected that will normally be reached during extreme cold weather. Outside design temperatures for the major cities in the United States are listed in Table 2-1.

*NOTE: Local weather authorities should be consulted for temperatures in localities other than those listed. Design temperatures for cities separated by only a few miles or at different altitudes may vary widely.*

**Design Temperature Difference.** The design temperature difference is the variation between the outside design temperature and the temperature desired inside the building. For example, if the rooms are to be heated to 70°F and the outside design temperature is 10° below zero, the design temperature difference is 80°.

*NOTE: If a room wall is adjacent to a partially heated space, such as a garage heated to 40°F, the garage temperature would be used as the outside design temperature in calculating the heat loss of that particular room wall.*

**Heat-Loss Determination.** List the design temperature difference on the tabulation sheet and multiply this difference by the "U" factor to determine the heat loss per square foot.

**Exposure.** "U" factors are calculated on a wind velocity of 15 mph. Higher velocities may be disregarded in determining the heat loss by transmission, but they must be taken into account when calculating the heat loss from infiltration.

**Basements.** Basements, unless used as living areas, normally have

### Table 2-1. Winter Outside Design Temperatures for Major Cities in the United States.

| State | City | Outside Design Temperature Commonly Used | State | City | Outside Design Temperature Commonly Used |
|-------|------|---|-------|------|---|
| Alabama | Birmingham | 10 | Missouri | Kansas City | −10 |
| Arizona | Tucson | 25 | | St. Louis | 0 |
| Arkansas | Little Rock | 5 | Montana | Billings | −25 |
| California | San Francisco | 35 | | Helena | −20 |
| | Los Angeles | 35 | | Miles City | −35 |
| Colorado | Denver | −10 | Nebraska | Lincoln | −10 |
| Connecticut | New Haven | 0 | | Valentine | −25 |
| Dist. of Columbia | Washington | 0 | Nevada | Reno | −5 |
| Florida | Jacksonville | 25 | New Hampshire | Concord | −15 |
| | Key West | 45 | New Jersey | Atlantic City | 5 |
| Georgia | Atlanta | 10 | | Trenton | 0 |
| | Savannah | 20 | New Mexico | Albuquerque | 0 |
| Idaho | Boise | −10 | New York | Albany | −90 |
| Illinois | Cairo | 0 | | Buffalo | −5 |
| | Chicago | −10 | | New York City | 0 |
| Indiana | Indianapolis | −10 | North Carolina | Asheville | 0 |
| Iowa | Des Moines | −15 | | Charlotte | 10 |
| | Sioux City | −20 | North Dakota | Bismarck | −30 |
| Kansas | Topeka | −10 | Ohio | Akron | 0 |
| Kentucky | Louisville | 0 | | Cincinnati | 0 |
| Louisiana | New Orleans | 20 | | Columbus | −10 |
| Maine | Portland | −5 | Oklahoma | Oklahoma City | 0 |
| Maryland | Baltimore | 0 | Oregon | Portland | 10 |
| Massachusetts | Boston | 0 | Pennsylvania | Erie | −5 |
| Michigan | Detroit | −10 | | Harrisburg | 0 |
| | Escanaba | −15 | | Philadelphia | 0 |
| | Sault Ste. Marie | −20 | Rhode Island | Providence | 0 |
| Minnesota | Duluth | −20 | South Carolina | Charleston | 15 |
| | Minneapolis | −20 | South Dakota | Huron | −20 |
| Mississippi | Vicksburg | 10 | Tennessee | Knoxville | 0 |

**Courtesy Dunham-Bush, Inc.**

**Table 2-1. Winter Outside Design Temperatures
for Major Cities in the United States (Cont'd)**

| State | City | Outside Design Temperature Commonly Used | State | City | Outside Design Temperature Commonly Used |
|-------|------|-------------------------------------------|-------|------|-------------------------------------------|
| Texas | Abilene | 15 | Washington | Seattle | 15 |
|  | Austin | 20 |  | Spokane | −15 |
|  | Brownsville | 30 | West Virginia | Parkersburg | −10 |
|  | Corpus Christi | 20 | Wisconsin | Green Bay | −20 |
|  | Houston | 20 |  | Madison | −15 |
|  | Dallas | 0 |  | Milwaukee | −15 |
| Utah | Salt Lake City | −10 | Wyoming | Cheyenne | −15 |
| Vermont | Burlington | −10 |  |  |  |
| Virginia | Lynchburg | 5 |  |  |  |
|  | Norfolk | 15 |  |  |  |

Courtesy Dunham-Bush, Inc.

enough piping to heat the space. If they are occupied, however, radiation should be installed. Heat loss should be calculated to determine the amount of radiation required.

Heat in a basement is generally lost by transmission and by infiltration through doors and windows. Basement walls and floors also have a heat loss, which is dependent on the ground temperature. The "U" factor for a basement wall below grade is but 0.10; therefore an assumed ground temperature will not greatly affect the tabulated transmission heat loss. Ground temperatures vary in accordance with the geographical location, however, so that the heat loss through the below-grade floor should be considered. A practical method is as follows:

Consider the basement wall as divided into two parts by a line at about the "mean" frost line. The wall above this line should be considered as being exposed to the outdoors, so the heat loss for this portion is calculated in the normal way. Everything below this line, including the basement floor, may be considered as being affected by a ground temperature selected from Table 2-2.

## AREA CALCULATION

### Step 2

After the transmission heat loss per square foot has been determined, it is necessary to find the total area of each surface.

61

## Table 2-2. Ground Temperatures Below the Frost Line

| State | City | Ground Temperature Commonly Used | State | City | Ground Temperature Commonly Used |
|---|---|---|---|---|---|
| Alabama | Birmingham | 66 | Nevada | Reno | 52 |
| Arizona | Tucson | 60 | New Hampshire | Concord | 47 |
| Arkansas | Little Rock | 65 | New Jersey | Atlantic City | 57 |
| California | San Francisco | 62 | New Mexico | Albuquerque | 57 |
|  | Los Angeles | 67 | New York | Albany | 48 |
| Colorado | Denver | 48 |  | New York City | 52 |
| Connecticut | New Haven | 52 | North Carolina | Greensboro | 62 |
| Dist. of Columbia | Washington | 57 | North Dakota | Bismarck | 42 |
| Florida | Jacksonville | 70 | Ohio | Cleveland | 52 |
|  | Key West | 78 |  | Cincinnati | 57 |
| Georgia | Atlanta | 65 | Oklahoma | Oklahoma City | 62 |
| Idaho | Boise | 52 | Oregon | Portland | 52 |
| Illinois | Cairo | 60 | Pennsylvania | Pittsburgh | 52 |
|  | Chicago | 52 |  | Philadelphia | 52 |
|  | Peoria | 55 | Rhode Island | Providence | 52 |
| Indiana | Indianapolis | 55 | South Carolina | Greenville | 67 |
| Iowa | Des Moines | 52 | South Dakota | Huron | 47 |
| Kentucky | Louisville | 57 | Tennessee | Knoxville | 61 |
| Louisiana | New Orleans | 72 | Texas | Abilene | 62 |
| Maine | Portland | 45 |  | Dallas | 67 |
| Maryland | Baltimore | 57 |  | Corpus Christi | 72 |
| Massachusetts | Boston | 48 | Utah | Salt Lake City | 52 |
| Michigan | Detroit | 48 | Vermont | Burlington | 46 |
| Minnesota | Duluth | 41 | Virginia | Richmond | 57 |
|  | Minneapolis | 44 | Washington | Seattle | 52 |
| Mississippi | Vicksburg | 67 | West Virginia | Parkersburg | 52 |
| Missouri | Kansas City | 57 | Wisconsin | Green Bay | 44 |
| Montana | Billings | 42 |  | Madison | 47 |
| Nebraska | Lincoln | 52 | Wyoming | Cheyenne | 42 |

Courtesy Dunham-Bush, Inc.

**Rooms.** List each room, using two lines for large rooms that have various sizes of windows and doors. List the length, width, and ceiling height of each room.

1. Determine the floor area by multiplying the room length by the width (Column 1 × 2, Fig. 2-2).
2. Calculate the ceiling area in the same manner as the floor.
3. Calculate the total outside wall area by multiplying the length and width of each room by the ceiling height (Column 1 × Column 3, and Column 2 × Column 3, Fig. 2-2).

**Windows and Doors.** List the number of windows of each type and their sizes. Window sizes are based on the outside measure-

| Room | 1 | 2 | 3 | 4 | 5 | 6 | 7 | 8 | 9 | 10 | 11 |
|---|---|---|---|---|---|---|---|---|---|---|---|
| | Room Size | | | Floor Area | Ceiling Area | Gross Wall Area | Windows and Doors | | | | Net Wall Area |
| | L | W | H | | | | No. | W | H | Area | |

**STEP 2—area calculations**

| Room | L | W | H | Floor Area | Ceiling Area | Gross Wall Area | No. | W | H | Area | Net Wall Area |
|---|---|---|---|---|---|---|---|---|---|---|---|
| Living rm. | 12 | 4 | $8^6$ | | 290 | 317 | D 1 | 3 | $6^8$ | 20 | 2 52 |
| | 20 | $11^4$ | $8^6$ | | | | W 1 | 9 | 5 | 45 | |
| | 10 | 16 | $9^6$ | | | | | | | | |
| dining rm. | $9^8$ | 9 | $8^6$ | | 87 | 159 | W3 | 3 | 5 | 45 | |
| kitchen | 7 | $8^6$ | $8^6$ | | 76 | 72 | D 1 | 3 | $6^8$ | 20 | |
| | $2^6$ | $6^6$ | $8^6$ | | | | | | | | |
| bath | 7 | 5 | $8^6$ | | 35 | 43 | W 1 | 3 | 3 | 6 | 37 |
| bedroom #1. | $13^4$ | $11^4$ | $8^6$ | | 151 | 210 | W3 | 3 | 5 | 45 | 165 |
| bedroom #2. | $10^8$ | $9^7$ | $8^6$ | | 102 | 172 | W2 | 3 | 5 | 30 | 142 |
| recreation rm. | 18 | 13 | $8^6$ | 234 | | | | | | | |
| wall above floor | | 31 | $3^6$ | | | 109 | W2 | 3 | $1^6$ | 9 | 100 |
| wall below floor | | 31 | $3^6$ | | | 109 | | | | | 109 |

① ceiling area includes area over stairway living rm. proper and area over bay window

Courtesy Dunham-Bush, Inc.

**Fig. 2-2. A sample tabulation sheet used to calculate area.**

ments of the sash. See Fig. 2-3. Calculate the area of each window. List the number of doors and their sizes. Door sizes are generally based on the actual size of the swinging door. Calculate the area of each door. Calculate the net wall area by subtracting the total area of the windows and doors (Column 10) from the gross wall area (Column 6).

## HEAT LOSS FROM TRANSMISSION

### Step 3

Calculate the heat loss through walls by multiplying the total wall area (last column in Fig. 2-2) by "HL" (last column in Fig.

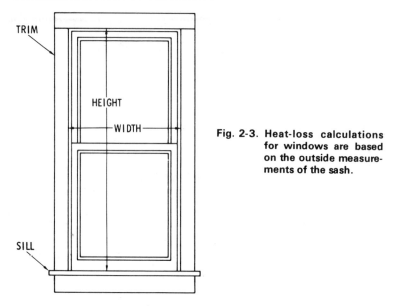

Fig. 2-3. Heat-loss calculations for windows are based on the outside measurements of the sash.

2-1). Calculate the heat loss through the floors by multiplying the floor area (Column 4, Fig. 2-2) by "HL." Calculate the heat loss through the ceilings by multiplying the ceiling area (Column 5, Fig. 2-2) by "HL." Finally, determine the heat loss through the windows and doors by multiplying the window and door areas (Column 10, Fig. 2-2) by "HL." These various losses are listed on a form similar to the one in Fig. 2-4.

## INFILTRATION HEAT LOSS

One of two methods may be used for calculating the heat loss due to infiltration around windows and doors—the *air-change method* or the *crack* method.

### Air-Change Method

The air-change method may be used with some accuracy for calculating heat loss of residential construction. The amount of air leakage is estimated by assuming a certain number of air changes per hour. This method does not consider velocities.

| | 12 | 13 | 14 | 15 | 16 | 17 | 18 | 19 |
|---|---|---|---|---|---|---|---|---|
| **STEP 3—heat loss from transmission** | | | | | | | | |
| Room | Walls | | Floors | | Ceilings (Roofs) | | Windows and Doors | |
| | 'HL' | Heat Loss | 'HL' | Heat Loss | 'HL' | Heat Loss | 'HL' | Heat Loss |
| *Living room* | 10.4 | 2624 | | | 9.8 | 2552 | 36.0 | 2340 |
| *dining room* | 10.4 | 1182 | | | 8.8 | 766 | 36.0 | 1620 |
| *Kitchen* | 10.4 | 544 | | | 8.8 | 667 | 36.0 | 720 |
| *bath* | 10.4 | 380 | | | 8.8 | 308 | 36.0 | 216 |
| *bedroom #1* | 10.4 | 1712 | | | 8.8 | 1330 | 36.0 | 1620 |
| *Bedroom #2* | 10.4 | 1478 | | | 8.8 | 899 | 36.0 | 1080 |
| *recreation room* | | | 1.8 | 421 | | | | |
| *Wall above grade* | 8.0 | 796 | | | 90.4 | 814 | | |
| *wall below grade* | 8.0 | 195 | | | | | | |
| | | | | | | | | |
| | | | | | | | | |
| | | | | | | | | |

Courtesy Dunham-Bush, Inc.

Fig. 2-4. Sample tabulation sheet used in calculating heat loss by transmission.

## Crack Method

The crack method is based on actual air leakage through the cracks around windows and doors according to wind velocities that may be expected in the particular area. See Table 2-3. The crack method is recommended as more accurate for determining the infiltration heat loss of a building.

Calculate the lineal feet of crack. The length of cracks for a double-hung window is three times the width plus two times the height. For a pivoted metal sash, the length of the crack is the total perimeter of the ventilating sections.

65

It is not necessary to use the total amount of crack when calculating infiltration heat loss because the wind does not blow through the cracks on all sides of a building at the same time. For instance, in a building without partitions, the air enters through cracks on the windward side and must leave through cracks on the opposite side. For this type of building, take ½ the total lineal feet of crack. For a room with only one exposed wall, use all the crack; but in no case take less than half of the crack. List the leakage per foot of crack (Table 2-3). This is the cubic feet of air per hour per foot of crack.

**Table 2-3. Typical Window Infiltration Chart**
(cubic feet per foot of crack per hour)

| Window Type | Remarks | Wind Velocity (miles per hour) | | | | | |
|---|---|---|---|---|---|---|---|
| | | 5 | 10 | 15 | 20 | 25 | 30 |
| Unlocked woodsash windows, double-hung | Perimeter of frame in wood-frame building | 2 | 5 | 10 | 16 | 22 | 28 |
| | Caulked perimeter in masonry wall | 1 | 3 | 4 | 5 | 6 | 7 |
| | Non-caulked frame perimeter in masonry wall | 2 | 3 | 12 | 18 | 25 | 34 |
| | Poorly fitted window, no weather-stripping with above hairline cracks including wood-frame leakage | 25 | 65 | 102 | 140 | 180 | 240 |
| Metal windows, double-hung | Locked without weatherstripping | 18 | 42 | 68 | 92 | 120 | 148 |
| | Unlocked without weatherstripping | 18 | 45 | 72 | 98 | 130 | 164 |
| | Unlocked but weatherstripped | 4 | 16 | 30 | 42 | 56 | 72 |

*NOTE: Room exposure should be considered in selecting wind velocities. Prevailing winter winds in most sections of the country are north, northwest, and west, and will have an effect on the rooms on these sides of the building.*

Crack thickness and the clearance determine the leakage through a double-hung window. The difference between the width of the window sash and the width of the track formed by the window guides is the clearance. When calculating the infiltration loss of a double-hung wood window, use the unlocked condition.

For a well-fitted door, use the leakage values for a poorly fitted window. Double this value if the door is poorly fitted; if the door is weatherstripped, use ½ this value. For a commercial building, where the door may be frequently opened, use 3 times this value. Calculate the infiltration heat loss by using the following formula:

$$L \times B \times \text{temp. diff.} \times .018 \times \text{Btu/hr infiltration heat loss}$$

where,

L is the lineal feet of crack (Column 25, Fig. 2-5).

B is the air leakage in cubic feet per hour per foot of crack (Column 26, Fig. 2-5).

| Room | 20 No. Exp. Walls | 21 No. Air Chgs. | 22 Room Vol-ume | 23 Design Temp. Diff. x .018 | 24 Heat Loss | 25 Lineal ft. of Crack | 26 Leak-age per Ft. (Cu. ft. /Hr.) | 27 Design Temp. Diff. x .018 | 28 Heat Loss | 29 Total Heat Loss† (Btu /hr.) |
|---|---|---|---|---|---|---|---|---|---|---|
| | | | | | STEP 4—heat loss from infiltration | | | | | |
| | | | AIR CHANGE METHOD | | | CRACK METHOD | | | | |
| *Living Room* | | _ | X_ | X_ | = _ | ᴰ19 | ×70ᴰ× | 1.44 | =1915 | 12,818 |
| | | _ | X_ | X_ | = _ | ᵂ48 | ×49ᴰ× | 1.44 | =3387 | |
| | | _ | X_ | X_ | = _ | X_ | X_ | = _ | | |
| *dining rm.* | | _ | X_ | X_ | = _ | 32 | ×24× | 1.44 | =1106 | 4674 |
| *kitchen* | | _ | X_ | X_ | = _ | 19 | ×56ᴰ× | 1.44 | =1532 | 3463 |
| | | _ | X_ | X_ | = _ | X_ | X_ | = _ | | |
| *bath* | | _ | X_ | X_ | = _ | 10 | ×24× | 1.44 | =346 | 1250 |
| *bedroom #1* | | _ | X_ | X_ | = _ | 32 | ×24× | 1.44 | =1106 | 5768 |
| *bedroom #2* | | _ | X_ | X_ | = _ | 16 | ×24× | 1.44 | =553 | 4010 |
| *recreation rm* | | _ | X_ | X_ | = _ | 18 | ×52× | 1.44 | =1348 | 3574 |
| | | _ | X_ | X_ | = _ | X_ | X_ | = _ | | |
| | | _ | X_ | X_ | = _ | ② The leakage of a poorly fitted window (double if exposition) | | | | |
| | | _ | X_ | X_ | = _ | ③ door leakage based on the leakage of a poorly fitted window plus 25%. | | | | |
| | | _ | X_ | X_ | = _ | Windows leakage based on that accuracy fitted windows plus 25% | | | | |
| | | _ | X_ | X_ | = _ | 25% allows for opening of doors and for fireplace. | | | | |

Fig. 2-5. A sample tabulation sheet used to calculate the heat loss of infiltration.

*Example:* What is the Btu/hr infiltration heat loss through the crack of a 3' × 5' poorly fitted window, nonweatherstripped, having a $\frac{3}{32}$" crack and a $\frac{3}{32}$" clearance? Wind velocity is expected to average 10 mph. Outside design temperature selected is 0°F and the inside air temperature will be 70°F.

*Answer:*According to Table 2-3, the air leakage through a window of this type will be 65 cubic feet per hour per foot of crack. Therefore, B equals 65. Lineal feet of crack (L) will be (2 × 5) plus (3 × 3), or 19 feet. The design temperature difference is 70°. Subtracting in the formula:

$$65 \times 19 \times 70 \times 0.018 = 1556 \text{ Btu/hr}$$

The total heat loss can now be determined by adding Columns 13, 15, 17, and 19 of Fig. 2-4, and Column 28 of Fig. 2-5. The total heat loss is placed in Column 29 of Fig. 2-5. After the heat-loss calculations have been computed, there are often factors to be considered. The heating capacity of the radiation placed in a room is (on the average) the same as that in Column 29 of Fig. 2-5. However, for some installations it will be necessary to make allowances for certain conditions after the heat loss has been calculated. For example, the radiation for bathrooms should be approximately 20% higher than the figured heat loss to make certain that comfortable conditions exist for bathing. High ceilings are another example. Since warm air is lighter than cool air, it rises and accumulates at the ceiling, resulting in a temperature differential between the ceiling and floor. The term applied to this condition is *stratification.* For ceilings 12' and under, this temperature difference is almost negligible. However, if the ceiling is higher, the difference is almost negligible. However, if the ceiling is higher, the difference in temperature may be considerable, depending on the type of heating units installed. Cast-iron radiators without shields, or a warm-air heating system, if not carefully designed, are most likely to cause a stratification condition. Unit heaters, if properly selected and installed, and convectors reduce stratification by causing air movement. A good solution is the installation of baseboard radiation, as the outlet air is warm, not hot, and the air flow is near the floor.

*NOTE: In practice, it is common to add 20% to the heating capacity of the units to be installed for each foot in ceiling height from 12' to 24'. If a ceiling is over 25' high, 25% can be added to the heating capacity of the units to be installed.*

Some buildings, such as churches, lodges, gymnasiums, and large meeting halls, may need to be heated only on certain days. In extremely cold weather, the building may become very cold, requiring extra heating to raise the room temperatures to a comfortable level in a short time. A common practice in this type of building is to add an approximate 25% to the radiation heating capacity if specific calculations are not made for a determination of the actual radiation requirements to provide a quick rise in temperature.

## EDR DETERMINATION

Hot-water system radiation to be installed is generally selected on the basis of the Btu/hr capacity of each unit. With steam, the radiation to be used is chosen on the basis of the square-feet EDR capacity of each unit. EDR (Equivalent Direct Radiation) originally was used in rating direct radiation (freestanding, cast-iron radiators, and pipe coils) on the basis of the square-foot area of the heating surface. Radiators, however, varied in heat output per square foot of surface depending on section depth, width, and height. As a result, a standard was brought into being—that which is now used—240 Btu/hr output per square foot of heating surface. This standard is in use for rating the radiation used with steam.

A square foot of EDR equals 240 Btu/hr. As an example: If it requires 14,400 Btu/hr to heat a room, the required steam radiation would be 60 square feet of EDR capacity.

## RADIATION SELECTION

### Convector Radiation

A well-designed convector (Fig. 2-6) supplies heat by a gentle air motion and a mild radiant effect for proper comfort condi-

HEAVY GAUGE CASING

KNOB-OPERATED DAMPER

ALL CORNERS ROUNDED

ELEMENT BRACKET

LOUVRE TYPE GRILL

CAST BRONZE HEADERS

HEAVY PROTECTING PLATES

NON-CORROSIVE FINS

Courtesy Dunham-Bush, Inc.

**Fig. 2-6. A convector radiator.**

tions. The absence of intense radiated heat, such as occurs with radiators, permits the occupants to be comfortable even when close to the convectors. The unit may be equipped with a damper, which regulates the air flow.

A convector is composed of two major parts—the heating element and the enclosure. The heating element consists of nonferrous fins attached to copper tubes. The size of the element, location of the outlet, and the enclosure height determine the convector capacity. The enclosure is actually a flue or chimney that directs air through the heating element and out into the room.

Factors to consider when selecting convectors are:

1. Limitations involved by the building construction.
2. Space available, and the cost and appearance of the convectors.
3. Interior decoration, occupancy, and type of building.

A convector should be located so as to offset any air currents that could result in discomfort. The total number of units required is determined by placing them according to exposure requirements. Taking into consideration the probable air currents, radiation subdivisions should be made so as to locate the convectors in high heat-loss areas. Distribute the radiation so as to maintain a uniform temperature of both the air and the wall surfaces. At the bottom of stairwells and under very high windows where a downdraft effect can be very pronounced, front outlet convectors are preferable to top outlet and sloping types. Fig. 2-7 shows a typical wall-hung convector for use with a hot-water system.

**Fig. 2-7.** **A typical wall-hung convector showing the suggested installation measurements.**

The piping layout and the type of cabinet determine the manner in which the convector heating elements are connected. The supply connections to the heating element may be made at the bottom, top, or end of the inlet header. The return connections are made at the bottom or end of the opposite header. Fig. 2-8 shows a typical convector connection, and Fig. 2-9 the typical piping connections for steam systems.

Piping connections for hot-water systems are illustrated in Fig. 2-10. An air chamber and vent should be part of each convector installation. If it is not desirable for the air valve to protrude through the side of the convector, the air chamber may be turned and a section of copper tubing connected between the air chamber and air valve. The valve can then be located below the heating element, near the inlet air opening, and will be easily accessible.

71

RADIATOR
VALVE

SUPPLY
STUB

FLOOR
PLATE

SUPPLY
RUNOUT

SUPPLY
RISER

Courtesy Dunham-Bush, Inc.

Fig. 2-8. Typical connection to a convector in a steam heating system.

## Baseboard Radiation

Baseboard radiation has the advantages of both convected and radiant heat (Fig. 2-11). Fig. 2-11 shows the cross section of a typical flush-back type, while Fig. 2-12 shows a recessed type.

A baseboard convector consists of two major parts—the baseboard enclosure and the heating element or elements. The enclosure is generally constructed of sheet steel with openings so placed as to be inconspicuous. These openings permit the air to pass over the heating element and into the space to be heated. The heating element is a finned tube usually made of copper with aluminum fins or steel with steel fins.

This type of radiation is generally located along the outside wall,

Courtesy Dunham-Bush, Inc.

**Fig. 2-9. Convector piping connections in a steam heating system.**

73

AIR CHAMBER

PIPE PLUG

AIR VALVE

HEADER

STREET
ELBOW

REGULATING
ELBOW

STREET ELBOW

Courtesy Dunham-Bush, Inc.

**Fig. 2-10. Convector piping connections in a hot-water heating system.**

where it may counteract the cooler air currents resulting from heat loss. Generally, with the ceiling-to-floor temperature is very uniform with this type of radiation, varying less than 4° as compared to the usual 12° with other types of radiation. Baseboard radiation is particularly recommended for rooms having wide windows with low sills.

## Fin-Tube Radiation

Fin-tube radiation (Fig. 2-13) is designed to meet the demand for a heating unit that is shallow in depth, low in height, attractive in appearance, and high in heating capacity. It is adaptable for almost any building, but is particularly useful in schools, industrial plants, and public and commercial buildings. A series of units may be used to make one long, attractive enclosure, and may be used with either hot water or steam.

This type of radiation should be mounted close to the floor (the height from the floor to the bottom of the cover should not be less than 4″ in order to permit an unrestricted air flow). Capacity tables in manufacturers' information folders and brochures are usually based on this position as the one that provides maximum distribution of heat. If, however, this type of radiation is mounted

FRONT SECTION OVERLAPS BACK
SECTION TO PROVIDE AN
EFFECTIVE AIR SEAL

DAMPER LATCH

DESIGN OF FRONT SECTION GUIDES
HEATED AIR AWAY FROM WALLS,
MINIMIZES SMUDGING

BACK SECTION SECURED
TO WALL. SERVES AS
SUPPORT FOR FRONT
SECTION

HEATED AIR IS CONVECTED
OUT INTO ROOM THROUGH
SERIES OF INCONSPICUOUS
OPENINGS

HEATING ELEMENT SUPPORTED
BY HANGER SECURED TO WALL

FRONT SECTION MAY BE
PAINTED TO SUIT YOUR
TASTE AFTER INSTALLATION

METAL FINS, ATTACHED
TO 1" TUBE, PROVIDE
FAST HEAT TRANSFER

ROOM AIR ENTERS THROUGH
SERIES OF INCONSPICUOUS
OPENINGS NEAR FLOOR

DAMPER CAN BE PROVIDED
FOR INDIVIDUAL ROOM
HEAT CONTROL

"BUMPER" KEEPS DIRT OUT
OF BASEBOARD, SPEEDS
ROOM CLEANING

BOTH BACK AND FRONT SECTIONS
MADE OF HEAVY GAUGE STEEL

Courtesy Dunham-Bush, Inc.

**Fig. 2-11. Cross section of a typical flashback baseboard radiator.**

too high, cold air may pocket at the floor level. Therefore, it
should never be mounted any higher than immediately below the
line of the windowsill, except for special jobs such as clerestory
windows, dormers, and skylights where it is used to eliminate
downdrafts.

Appearance is generally a controlling factor in the selection of
radiation. The tiered type shown in Fig. 2-14 is often used in
schools and other public buildings. The front outlet in the cover
avoids the danger of smudging the walls. When a front outlet
might be blocked by furniture or equipment, a sloping top or

FRONT SECTION OVERLAPS BACK SECTION TO PROVIDE AN EFFECTIVE AIR SEAL

DAMPER LATCH

DESIGN OF FRONT SECTION GUIDES HEATED AIR AWAY FROM WALLS, MINIMIZES SMUDGING

BACK SECTION SECURED TO WALL. SERVES AS SUPPORT FOR FRONT SECTION

HEATED AIR IS CONVECTED OUT INTO ROOM THROUGH SERIES OF INCONSPICUOUS OPENINGS

HEATING ELEMENT SUPPORTED BY HANGER SECURED TO WALL

FRONT SECTION MAY BE PAINTED TO SUIT YOUR TASTE AFTER INSTALLATION

METAL FINS, ATTACHED TO 1" TUBE, PROVIDE FAST HEAT TRANSFER

ROOM AIR ENTERS THROUGH SERIES OF INCONSPICUOUS OPENINGS NEAR FLOOR

DAMPER CAN BE PROVIDED FOR INDIVIDUAL ROOM HEAT CONTROL

"BUMPER" KEEPS DIRT OUT OF BASEBOARD SPEEDS ROOM CLEANING

BOTH BACK AND FRONT SECTIONS MADE OF HEAVY GAUGE STEEL

Courtesy Dunham-Bush, Inc.

Fig. 2-12. Cross section of a typical reversed baseboard radiator.

flat-topped cover should be used. In applications where the covers may be damaged, replaceable covers such as the flat-topped or expanded-metal type may be preferable.

The wall construction and the extent of cold drafts may be a deciding factor on the kind of cover that should be installed. On some masonry walls, for example, the lack of a nailing strip may make the mounting of covers that fasten to the walls costly. Backs may be necessary in some instances to prevent leakage of air into furred spaces. In other cases, it may be necessary to place an insulating board in back of the units.

Fig. 2-13. Typical covers for fin-tube radiators.

Fig. 2-14. Cross section of a three-tier fin-tube wall radiator.

## BOILER AND PUMP SIZES

After all heating calculations have been tabulated, and after the type and size of radiation and boiler have been determined, the pumps and/or units can be selected from manufacturers' literature.

CHAPTER 3

# Steam Heating Systems

The widespread use of steam for space heating points up the long-recognized fact that steam, as a heating medium, has numerous characteristics that can be employed advantageously. Some of the most important advantages are discussed below.

*Steam's ability to give off heat.* Properties of saturated steam are shown in steam tables and give much information about the temperature and heat (energy) contained in one pound of steam for any pressure. As an example, to change one pound of water from 212°F into steam at the same temperature of 212°F at atmospheric pressure (14.7 psi) requires a heat content of 1150.4 Btu, which is made up of 180.1 Btu of *sensible heat* (the heat required to raise one pound of water from 32°F to 212°F) and 970.3 Btu of *latent heat.* The latent heat is the heat added to change the one pound of water from 212°F into steam at 212°. This stored up latent heat is required to transform the water into steam, and it reappears as heat when the process is reversed to condense the steam into water.

Because of this basic fact, the high latent heat of vaporization of a pound of steam permits a large quantity of heat to be transmitted efficiently from the boiler to the heating unit or radiator with little change in temperature.

*Steam promotes its own circulation through piping.* Steam will flow naturally from a higher pressure as generated in the boiler to a lower pressure (existing in the steam mains). Circulation or flow is caused by the lowering of the steam pressure along the steam supply mains and in the heating units due to pipe friction and to the condensing process of steam as it gives up heat to the space being heated. This action is shown in Fig. 3-1. Because of this, the natural flow of steam does not require a pump such as needed for hot-water heating, or a fan as employed in warm-air heating.

*Steam heats more readily.* Steam circulates through a heating system faster than other fluid mediums. This can be important where fast pickup of the space temperature is desired. It will also

Fig. 3-1. Steam condensing as it gives up heat to the space being heated.

cool down more rapidly when circulation is stopped. This is an important consideration in spring and fall when comfort conditions can be adversely affected by long heating-up or slow cooling-down periods.

*Flexibility of steam heating.* Other advantages in using steam as a heating medium can be found in the easy adaptabiliy to meet unusual conditions of heat requirement with a minimum of attention and maintenance. Here are some examples:

1. Temporary heat during construction is easily provided without undue risks and danger of freeze-ups.
2. Additional heating units can be added to existing systems without making basic changes in the system design.
3. Increased heat output from heating units can be easily accomplished by increasing the steam pressure the proper amount.
4. Steam heating systems are not prone to leak; however, leaks that may occur in the system piping, pipe fittings, or equipment cause less damage than leaks in a system using hot water. A cubic foot of steam condenses into a relatively small quantity of water. In many cases, a small leak does not cause any accumulation of water at the location of the leak; instead, it evaporates into the air and causes no damage.
5. Repair or replacement of system components such as valves, traps, heating units (radiators), and similar equipment can be made by simply closing off the steam supply. It is not necessary to drain the system and to spend additional time to reestablish circulation. There is less need to worry about freezing since the water in a steam heating system is mainly in the boiler. Boiler water can easily be protected from freezing during shutdowns, or during new construction, repairs, or replacement of parts, by installing an aquastat below the boiler water level to control water temperature.
6. Steam is a flexible medium when used in combination processes and heating applications. These often require different pressures, which are easily obtained. In addition, exhaust steam, when available, can be utilized to the fullest advantage.
7. Steam heating systems are considered to be "lifetime"

81

investments. Many highly efficient systems are in operation today after more than 50 years of service. Fig. 3-2 shows a steam boiler with the water at normal operating level, and a hot-water boiler with the system completely filled with water.

BOILER
WATER LEVEL

STEAM BOILER
WATER AT BOILER WATER LEVEL ONLY

HOT WATER BOILER
SYSTEM COMPLETELY FILLED WITH WATER

**Fig. 3-2. Water content in boiler.**

*Steam is easy to distribute and control in a heating system.* The distribution of steam to heating units or radiators is easy to accomplish with distribution orifices located at the steam inlet to the unit. Metering orifices an also be employed when used with proper controls to maintain steam pressure in accordance with the flow characteristic of the metering orifices. These orifices can be either fixed or a variable type. Controls for steam systems are simple and effective. They include those used to control space temperatures by the application of ON-OFF valves. Modulating controls can also be applied that respond to indoor-outdoor temperature conditions to control the quantity of steam flowing to orificed radiators.

*Where steam can be used.* Steam can be used as the heating medium for all types of heating units such as convectors, wall fin-tube radiation, cast-iron radiators, unit heaters, unit ventila-tors, heating and ventilating units, all types of coils in ventilating and air-conditioning systems and steam absorption units used in air conditioning. Steam can be used for all types of systems applicable to a variety of building designs ranging from residences to large industrial, commercial, multistory apartment groups, offices, churches, or schools.

*Recommended steam applications.* The following applications for the use of steam are recommended to provide troublefree and efficient heating systems:

1. Where there is more than one job to do, such as providing comfort heating as well as steam for processes in:
   Industrial plants.
   Hospitals.
   Institutions.
   Restaurants.
   Dry cleaning plants.
   Laundries.
2. Where outdoor air is heated for ventilation (especially in cold climates) as in:
   Factories.
   School classrooms.
   Auditoriums.
   Buildings with central air conditioning or ventilating systems.
   Gymnasiums.
3. Where there is a surplus of steam from processes that can be used for air cooling or water chilling.
4. Where the heating medium must travel a great distance from the boiler to the heating units, as found in:
   Multistory buildings.
   Long, rambling buildings.
   Scattered buildings.
   Supplied from a central station.
5. Where intermittent changes in heat loads are required, as in:
   Schools.
   Office buildings.
   Churches.
6. Where central heat control or individual room control of temperature is important, as found in:
   Schools.
   Hospitals.

Office buildings.

Hotels and motels.

7. Where there is a chance of freezing in cold climates or where subfreezing air is handled.

8. Where there may be additions or alterations of space or change in occupancy in a building.

9. Where extra heat is needed as in buildings with large or frequently used doors, as in:

Department stores.

Garages.

Shipping departments or warehouses.

Airplane hangars.

# ONE-PIPE STEAM HEATING SYSTEMS

All steam heating systems can be said to be of two basic types: either one-pipe systems or two-pipe systems. The one-pipe system uses the same pipe to deliver steam to a unit and return the condensate from the unit. The unit (radiator, unit heater, etc.) has only one piping connection; the steam enters the unit and the condensate returns through the one piping connection at the same time.

Correctly designed and installed one-pipe systems require a minimum of mechanical equipment. Open-type air vents (shown in Figs. 3-3A and B) are used on the end of mains and on each heating unit. If the system is correctly designed and installed, the condensate will flow (return) back to the boiler by gravity. The advantage of the one-pipe system is that it is very dependable and the initial cost is low. Fig. 3-4 shows a typical modern one-pipe installation of a simple up-feed gravity system.

## The Steam Boiler

The boiler in most cases will be fired automatically and will be equipped with controls for maintaining the system pressure. It should be equipped with an automatic water feeder and a low-water cutoff. It should have the required safety devices for proper combustion of the fuel used.

## The Steam-Supply Main

The steam-supply main conducts the steam from the boiler to the various radiators in the system. Condensate flows back from the radiators to the drip connection through the steam-supply

(A) An open-type radiator air vent.

(B) An end-of-main vent.

**Fig. 3-3. Typical one-pipe system vents.**

**Fig. 3-4. A basic one-pipe up-feed gravity steam system.**

85

main. As shown in Fig. 3-4, the condensate flow is in the same direction as the steam flow. This system is called a parallel flow system.

## Radiator Valves

The steam enters each radiator through an angle-pattern radiator supply valve installed at the bottom inlet tapping of the radiator. The valve should be an angle-pattern valve because steam must enter the heating unit and condensate must leave the unit through the same valve and at the same time. A straightway horizontal-pattern valve should not be used because it has a natural obstruction to the free flow of steam and condensate. A radiator supply valve on a single-pipe system can only by used to turn the radiator OFF completely or ON completely. It cannot be used as a throttling valve due to the noise that would be created. Furthermore, if a number of valves in a system were throttled, steam condensate could not readily return to the boiler. The correct valve for a one-pipe steam system is shown in Fig. 3-5.

## The Hartford Loop

The Hartford loop is a pressure-balancing loop that introduces full boiler pressure on the return side of the boiler. This pressure prevents *reversed circulation.* or water leaving the boiler via the return piping. Some engineers in the past have attempted to stop reversed circulation by installing a check valve in the return piping. This method is unsatisfactory, indeed unsafe, because dirt, corrosion, or scale can lodge under the check valve and prevent it from closing, permitting reversed circulation to occur. The Hartford loop is the only safe method of preventing reversed circulation. It is actually two loops, the first is an equalizer line and the second is an extension from the steam main, through the return and back to a connection at the equalizer loop, as shown in Fig. 3-4.

In order for the Hartford loop to be effective, the loop must be at least the full size of the return main, and the horizontal nipple at the point where the return connects should be 2″ below the boiler water line. It is important to use a close nipple to construct the

**Fig. 3-5. A typical radiator supply valve.**

Hartford loop; if a long nipple is used at this point and the water line of the boiler becomes low, water hammer noise will occur.

## End-of-Main Vent Valves

Air must be eliminated from the piping and radiation of a one-pipe steam system before steam can enter the piping or the radiators. Air that is present in the piping or in the radiators will block the flow of steam. An end-of-main vent should be installed on large one-pipe systems to assure quick venting of air from the horizontal steam main. Venting of the air will open a path and assure prompt distribution of steam to vertical runs and risers supplying steam to radiators. An end-of-main vent is made with a single port and has a much larger venting capacity than a radiator vent. A nonvacuum or open main vent is shown in Fig. 3-6. Main vents are never constructed for adjustable, variable venting because once all the air has been expelled form the steam main, the vent port is closed and has no further function until the next boiler firing or heating-up cycle occurs.

A vacuum end-of-main vent is shown in Fig. 3-7. It has all the features of a smaller vacuum-type radiator vent. A vacuum-type vent has all the features of an ordinary air vent, plus it prevents the return of air into the system through the vent valve.

End-of-main vents should always be installed near the end of

Fig. 3-6. A float-type, end-of-main, air-vent valve.

Fig. 3-7. A float-type, vacuum, end-of-main, air-vent valve.

the steam main. It is *not* good practice to install a main vent on the last fitting at the end of the steam main.

The correct way to install the end-of-main vent is shown in Fig. 3-8. When this vent is incorrectly installed (shown in Fig. 3-9), the float can be damaged by water surge, which can create a very high pressure. Although the surge pressure lasts only a fraction of a second, it can cause the float to collapse.

**Fig. 3-8. Correct way to install an end-of-main air vent.**

MAIN VENT

¾" COUPLING

6" TO 10"

DRY RETURN

AT LEAST 15"

CONNECT TO WET RETURN

Courtesy Hoffman Specialty ITT

**Fig. 3-9. Incorrect installation of an end-of-main air vent.**

MAIN VENT

STEAM MAIN

INCORRECT INSTALLATION

Courtesy Hoffman Specialty ITT

## Radiator Vent Valves

Air that is present in radiators will prevent the steam in the supply main from entering the radiator. Each radiator must be equipped with an air vent valve installed in the vent tapping located on the end opposite the supply valve. Radiator vent valves are made in two basic types:

89

1. The nonvacuum or open vent is available with a single nonadjustable port or with an adjustable port used for proportional venting.
2. The vacuum type has an adjustable port for proportional venting.

The functions of a vent valve in a heating system are:

1. To permit air to be vented (pushed out) so that the steam can occupy the space in the piping and heating units.
2. To close when steam contacts the valve, thus preventing the steam from escaping through the vent port.
3. To close when water contacts the valve and thus prevent the water from escaping through the vent port.
4. To reopen when the steam temperature has dropped sufficiently, or the water has drained away, thus permitting the air venting process to continue.
5. Vacuum vent *only*. An additional feature to prevent the return of air to the system.

## Proportional Venting

When the steam starts through a piping system that has more than one radiator, it is desirable for the steam to be able to enter all radiators at approximately the same time. Because there will be a difference in size (EDR—Equivalent Direct Radiation) of the radiators, due to the space being heated by each unit, single-port vent valves should not be used. Instead, adjustable-port vent valves should be installed on each unit of radiation and the vents adjusted to the size of each unit. When the vents are properly adjusted, steam will enter each unit of radiation at the same time during each firing cycle. An adjustable-port vent valve is shown in Fig. 3-10. A single-port (nonadjustable) vent valve is shown in Fig. 3-11.

We have followed the path of the steam through the supply main to each unit of radiation and to the end of the supply main. At the end of the supply main the steam will revert to the water from which it came and will become condensate. This condensate will be joined by the condensate that formed in the various units of radiation and will return to the boiler through the wet-return piping of the system. Referring again to Fig. 3-4, the section of piping between the end of the supply main and the main vent is called the *dry return*. The dry return is that portion of the return main located above the boiler water line. In addition to carrying

SIX PORTS

ADJUSTABLE PORT

VACUUM CHECK

SEAT

CONNECTION SIZE
¼" STRAIGHT SHANK

OPERATING PRESSURE
OF UP TO 1 ½" PSI

MAXIMUM PRESSURE
10 PSI

FLOAT

⅛" PIPE
THREAD

TONGUE

Courtesy Hoffman Specialty ITT

Fig. 3-10. An adjustable-port vent valve.

Fig. 3-11. A single-port (nonadjustable) air valve.

Courtesy Hoffman Specialty ITT

91

condensate, it also carries steam and air. The end of the dry return must be located at the proper height to maintain the minimum required distance for dimension A above the water line of the boiler.

Dimension A is shown in Fig. 3-4 and is the location of the end of the steam supply main above the boiler water line. *A small system having a total heat loss of not more than 100,000 Btu/hr* is sized on the basis of ⅛ psi. The three distances shown in Fig. 3-12, the pressure drop, the static head, and the safety factor, will be:

Pressure drop of system (⅛ psi)  =  3½″ of water

Static head (friction of wet return)  =  3½″ of water

Safety factor (twice the static head)  =  7″ of water

Total distance  =  14″ of water

For a small system, it is standard practice to make the minimum distance for dimension A not less than 18″.

For a larger system, assume that the piping is sized for a total pressure drop of ½ psi. The three distances would then be:

Fig. 3-12. Pressure drop in a one-pipe steam system.

| | |
|---|---|
| Pressure drop of system (½ psi) | = 14″ of water |
| Static head (friction of wet return) | = 4″ of water |
| Safety factor (twice the static head) | = 8″ of water |
| Total distance | = 26″ of water |

It is standard practice for a system based on ½ psi pressure drop to make the minimum distance for dimension A not less than 28″.

## The Wet Return

The portion of the return piping that carries the condensation back to the boiler and is installed below the level of the boiler water line is called the wet return. It is completely filled with water and does not carry air or steam. Since water will seek its own level when the system is first filled or is cold, the water in the wet return will be at the same level as the water in the boiler.

## Types of One-Pipe Systems

There are several one-pipe heating systems, determined by the piping arrangement. In a gravity system the condensate returns directly to the boiler. The proper slope or pitch must be given to the steam supply and dry-return mains in a gravity system in order to ensure the proper flow of steam, air, and condensate. The proper pitch is not less than 1″ in 20′ in the direction of the gravity flow of the condensate. The wet return requires no pitch. Some typical examples of different piping arrangements and types are shown and explained in the following paragraphs.

**The Counterflow System.** In a counterflow system the steam flows in the opposite direction to the flow of condensate. This system is limited to small residences, especially those with unexcavated basements. The sizing of the steam main depends on the load requirements; it must be one pipe size larger than the steam main required on other one-pipe systems, and it must pitch upward from the boiler toward the end of the main. A slope of 1″ in 10′ is required to facilitate proper condensate return to the boiler. The end of the steam supply main, dimension A in Fig. 3-12, must be of sufficient height above the boiler water line to return the condensate to the boiler. A counterflow system is shown in Fig. 3-13.

**Parallel Flow Systems.** In a parallel flow system, steam and condensate flow in the same direction in the horizontal steam or return mains. A parallel flow system with a wet return is shown in Fig. 3-14. A parallel flow system using a dry return is shown in Fig. 3-15. The end of the dry-return main drops below the boiler water line and becomes a wet return. In both the dry- and wet-return systems

Courtesy Hoffman Specialty ITT

Fig. 3-13. A counterflow one-pipe steam system.

Courtesy Hoffman Specialty ITT

Fig. 3-14. A parallel flow system using a wet return.

94

the condensate returns to the boiler through a Hartford loop. In both the dry- and wet-return systems, dimension A in Fig. 3-12 must be maintained.

**Parallel Flow Up-Feed Systems.** The parallel flow up-feed system is designed for buildings having more than one floor. As shown in Fig. 3-16, steam is distributed to the units of radiation through a basement main from which two up-feed risers supply the second

Courtesy Hoffman Specialty ITT

**Fig. 3-15. A parallel flow system using a dry return.**

Courtesy Hoffman Specialty ITT

**Fig. 3-16. A parallel flow up-feed system.**

and third floors. The up-feed risers are dripped back to the wet return. Connecting the risers, as shown in Fig. 3-17, is advisable in order to keep the horizontal main free of condensate accumulation and assure unobstructed flow of steam. The up-feed branch connection to the first-floor radiators are not dripped. The details of these connections are shown in Fig. 3-18. If the up-feed runout to a riser is not to be dripped to the wet return, the horizontal pipe must be increased one pipe size and a greater pitch provided, as shown in Fig. 3-19. Two methods of connecting a branch or runout to an up-feed riser are shown in Fig. 3-20. The 45° connection provides less obstruction to the free flow of steam to the riser, which must also carry the reverse flow of condensate from the riser.

**Parallel Flow Down-Feed System.** A one-pipe system with the distribution main installed above the radiators, as shown in Fig. 3-21, is called a down-feed system. In this system the steam and condensate flow the same way in the down-feed risers. The main steam supply riser should be installed directly from the boiler to the overhead supply main as shown in the illustration. The down-feed runout connections are taken from the bottom of the horizontal supply main, thus assuring the least accumulation of condensate in the main. The piping details for these runouts from the bottom of the main are shown in Fig. 3-22. If the main vent is installed on the end of the horizontal main, only one main vent will be required. Main vents may be installed on the risers (optional) as shown in Fig. 3-21; if they are installed on the riser, they must be located so that the dimension A (Fig. 3-12) is maintained at the proper height above the boiler water line, to prevent the vent valve from being closed by water rising to a sufficient height to raise the vent valve float.

### Returning the Condensate to the Boiler by Mechanical Means

A condensate pump must be used to return the condensate to the boiler when there is insufficient height to maintain dimension A at its minimum above the boiler water line.

When a condensate pump is used, the boiler pressure, the end of the steam main pressure, or the boiler water-line elevation have no bearing on the height of the end of the steam main as long as it is

Courtesy Hoffman Specialty ITT

Fig. 3-17. Correct installation of up-feed risers in a parallel flow up-feed system.

Courtesy Hoffman Specialty ITT

Fig. 3-18. Correct installation of an up-feed branch connection not dripped to the wet return.

Courtesy Hoffman Specialty ITT

Fig. 3-19. Correct installation of an up-feed riser not dripped to the wet return.

ACCEPTED

PREFERRED

45°

METHOD OF TAKING BRANCHES FROM MAIN

Courtesy Hoffman Specialty ITT

**Fig. 3-20. Preferred methods of taking branch connections from a main.**

SUPPLY MAIN

AIR VENT

VENT VALVE

SUPPLY RISER

DROP RISER

A

VENT (OPTIONAL)

WET RETURN

Courtesy Hoffman Specialty ITT

**Fig. 3-21. A parallel flow down-feed system.**

STEAM MAIN

PITCH ½″ in 10′

DOWN-FEED RISER

ELEVATION

END VIEW

Courtesy Hoffman Specialty ITT

**Fig. 3-22. All down-feed runout connections should be taken from the bottom of the main.**

above the maximum water level in the condensate pump receiver. The return main begins at the discharge of a float and thermostatic trap as shown in Fig. 3-22. This trap is sized to handle the maximum condensate load of the system if it is a single main, or the connected load of each individual main. It functions to discharge air and condensate accumulations into the return, and to close against the passage of steam. The return main for this system is referred to as a "no pressure return" because the open vent on the condensate pump receiver maintains it at atmospheric pressure. The piping must be uniformly pitched to the pump receiver without pockets, which will trap air and prevent the gravity flow of condensate to the receiver. The pump discharge is connected directly to the boiler return opening *without* the use of a Hartford loop. A Hartford loop connection can cause noise when used with a pumped discharge of condensate. It is not good practice to install the discharge from a float and thermostatic trap directly into, or close to, the pump receiver because the "flash" of steam vapor may be sufficient to affect the pump operation. "Flash," as used here, means the rapid passing into steam, of water at a high temperature when the pressure it is under is reduced. For best pump operation, the condensate temperature in the receiver should not exceed 200°; for this reason, the return piping should not be insulated when a pump is used to return the condensate to the boiler. Often construction conditions exist that will not permit the condensate to be returned by gravity flow to the condensate receiver if the receiver is installed above the floor, as shown in Fig. 3-23. In this case an underground receiver can be installed. The return main can also be installed underground to permit the condensate to return to the receiving tank by gravity flow. The condensate pump could then be mounted on the receiver, above ground, to pump the condensate back to the boiler. An underground receiver with the pump mounted on top is shown in Fig. 3-24.

When a horizontal main is installed without uniform pitch, usually due to improper or insufficient support, it can cause a water pocket resulting from a sag in the piping. Fig. 3-25 shows such a condition, and it is evident that a free flow of steam and air is impossible until the condition is corrected. Such a sag can cause noise when steam reaches the water pocket on a cold start. Also,

Courtesy Hoffman Specialty ITT

**Fig. 3-23. An above-ground condensate receiver and pump.**

Courtesy Hoffman Specialty ITT

**Fig. 3-24. A condensate receiver installed underground. The condensate pump is mounted on top of the receiver.**

Courtesy Hoffman Specialty ITT

**Fig. 3-25. The cause of noise in a sagging pipe.**

100

depending on its location, it can be responsible for water hammer noise and, sometimes, destructive damage to valves and vents.

There are times when it is necessary to install a steam main around a construction obstruction such a a steel beam or girder. The details of the proper method of piping around such an obstruction are shown in Fig. 3-26. Steam and air will flow above the obstruction, and the condensate will flow below.

SAME SIZE AS STEAM MAIN

PITCH

MINIMUM 1"

WET-RETURN MAIN SIZE

Courtesy Hoffman Specialty ITT

**Fig. 3-26. Correct method of piping around a beam.**

Another obstruction is a doorway that interferes with the uniform pitch of a dry return that must be located above the floor. Fig. 3-27 shows the correct method of piping around the doorway. When the piping is installed as shown, the condensate will flow in the piping below the doorway, and air will flow through the air loop over the top.

A radiator that does not set level because of a sagging floor, as shown in Fig. 3-28, can be the cause of noise and poor heating. Condensate remains trapped in the lower end of the radiator, thus setting up an obstruction to the steam which is trying to enter the radiator. Water hammer can result from this condition. The radiator should be set as level as possible. On a one-pipe system, a slight pitch of the radiator towards the supply end will help eliminate problems.

There are times when it may be desirable to add another radia-

AIR LOOP

PITCH

AT LEAST 1"

DRY RETURN    AIR AND CONDENSATE    CONDENSATE    PLUG IN TEE FOR CLEANOUT

Courtesy Hoffman Specialty ITT

**Fig. 3-27. The correct method for piping steam and condensate around a doorway.**

TRAPPED CONDENSATE

LEVEL FLOOR LINE

SAGGING FLOOR

Courtesy Hoffman Specialty ITT

**Fig. 3-28. The cause of noise in a radiator.**

tor to an existing one-pipe steam heating system. Fig. 3-29 shows the wrong and right methods for adding the additional radiator. Incorrect installation will cause the supply pipe from the existing radiator to fill with condensate and thus block the passage of steam to the added radiator. The right method allows the steam

from the supply pipe to go directly to the additional radiator. The
new steam-supply piping must be pitched as shown in Fig. 3-29 so
that condensate can flow back to the return system.

WRONG METHOD

ADDED RADIATOR

EXISTING RADIATOR

STEAM SUPPLY

RIGHT METHOD

ADDED RADIATOR

EXISTING RADIATOR

PITCH

STEAM SUPPLY

RADIATOR PEDESTALS
TO ENSURE PROPER
DRAINAGE OF CONDENSATE

PITCH PIE DOWN 1" FOR EACH
10' FOR PROPER DRAINAGE TO
BOILER FOR RETURN OF CONDENSATE

PIPE SIZE
TO SUIT LOAD

**Courtesy Hoffman Specialty ITT**

Fig. 3-29. The correct way to add a radiator to a system.

## Piping Fundamentals for One-Pipe Systems

Essential factors must be considered in the design and installation of the piping for a one-pipe steam system. To ensure the
proper flow of steam, air, and condensate, the steam-supply main
and the dry return must be installed with a uniform pitch. Fig. 3-30
shows the proper method of reducing the size of a pipe installed
horizontally. Use an eccentric reducing coupling instead of a

WATER POCKET

RIGHT
ECCENTRIC REDUCING COUPLING

WRONG
REGULAR REDUCING COUPLING

**Courtesy Hoffman Specialty ITT**

Fig. 3-30. How to reduce horizontal piping.

103

regular (concentric) reducing coupling. The use of the eccentric reducing coupling permits the continuance of uniform pitch without forming a water pocket, which restricts flow.

# TWO-PIPE STEAM HEATING SYSTEMS

A two-pipe heating system is a system in which the heating units have two piping connections, one used for the steam supply and the other used for the condensate return. Some of the same components used in one-pipe systems are used in two-pipe systems. Two-pipe systems are designed to operate at pressures ranging from subatmospheric (vacuum) to high pressure. Although they use many practical piping arrangements to provide up-flow or down-flow systems, they are conveniently classified by the method of returning the condensate to the boiler. Condensate can be returned to the boiler by gravity or by use of any one of several mechanical return means.

### The Steam Boiler

Boilers are manufactured as either cast-iron sectional boilers or steel boilers. The modern boiler is automatically fired, using coal, oil, or gas as the fuel. Suitable pressure controls, safety firing devices, and low-water cutoff valves should be installed on the boiler. Package boilers, furnished complete with all equipment factory assembled and tested, are used extensively in larger installations.

### Heating Units

Heating units such as cat-iron radiation, wall fin-tube radiation, convectors, unit ventilators, cabinet heaters, and unit heaters are used in two-pipe heating systems.

### Radiator Supply Valves

The supply valve for a radiator in a two-pipe system is installed at the inlet connection. The supply valves are globe valves and are made in angle, straight-way, and right- and left-hand patterns. The

104

angle-pattern valve is the most widely used, being adaptable to most heating unit installations. Radiator supply valves can be obtained in either the modulating (adjustable flow) or nonmodulating type. A supply valve that uses stem packing is shown in Fig. 3-31. A packless supply valve is shown in Fig. 3-32. The packless valve is a modulating valve also; leakage around the stem is prevented by the metal bellows. A packless valve is particularly suited for use in two-pipe vacuum systems since the construction of this valve prevents air leakage into the system and steam leakage from the system.

Courtesy Hoffman Specialty ITT

Fig. 3-31. A modulating supply valve with a spring-packed stem.

HANDLE

DIAPHRAGM

LEVERS, PLUNGERS,
AND STEM
ASSEMBLY

Courtesy Hoffman Specialty ITT

Fig. 3-32. A packless modulating supply valve.

105

## Thermostatic Steam Traps

Thermostatic steam traps are the most commonly used traps in two-pipe steam heating systems. They are of simple construction, small in physical size and weight with ample capacity for the usual heating system pressure. Thermostatic traps open in response to pressure as well as temperature to discharge condensate and air. When steam reaches the thermostatic element, the trap closes to prevent the passage of steam into the return.

Thermostatic traps are made in angle, swivel, straight-way, and vertical patterns. Thermostatic traps are installed on the return connection of a radiator, but they have many other uses also. They are used as drip traps and to handle the condensate from fan-coil units such a unit heaters. Two thermostatic traps are shown in Fig. 3-33. The cutaway views show the operating elements of these traps. The thermostatic elements and the seats are replaceable.

Fig. 3-33. Cutaway view of two thermostatic traps.

Courtesy Hoffman Specialty ITT

Some typical applications for thermostatic traps are shown in Fig. 3-34. Note that certain installations of thermostatic traps require the installation of a cooling leg to permit the condensate accumulation to cool sufficiently to open the trap and discharge the conden-

Fig. 3-34. Typical trap installations showing the use of a cooling leg.

sate into the return. An installation requiring the use of a cooling leg is shown in Fig. 3-34.

## Mechanical Steam Traps

Some of the mechanical traps used in two-pipe steam heating systems are: float and thermostatic traps, float traps, inverted bucket traps, and open or upright bucket traps. The different operating characteristics of these traps makes them applicable to a variety of steam main drips, riser drips, and heating units.

**Float and Thermostatic Traps.** A float and thermostatic trap, called an F&T trap, is opened by the action of the condensate collecting in the trap and raising the float. As the float raises, it opens the discharge port and permits the condensate to enter the return piping. The float causes a throttling action of the valve in the seat port; therefore, the discharge of the condensate is continuous. When air or condensate is present in the trap at a temperature below its designed closing pressure, the thermostatic air bypass remains open. The thermostatic air bypass closes as steam enters the trap. Cooling legs are not required with an F&T trap installation because the trap will discharge condensate at any temperature up to a temperature very close to the saturated steam

107

temperature corresponding to the pressure at the trap inlet. An F&T trap is shown in Fig. 3-35. F&T traps are used on unit heaters, unit ventilators, to drip the ends of steam mains, and to drip the heels of up-feed steam risers and down-feed steam risers.

AIR PASSAGE   THERMOSTAT   FLOAT BALL   INLET

FULCRUM   OUTLET   SEAT   PIN   BODY

Courtesy Hoffman Specialty ITT

**Fig. 3-35. The action of a float and thermostatic trap.**

**Inverted Bucket Trap.** Bucket traps are used to drain condensate and air from equipment such as unit heaters, steam coils, steam heated vats, and pressing machinery, and to lift condensate to overhead return mains. On a new installation the trap must be primed by pouring water through the plug hole on top of the trap. The operation of an inverted bucket trap is shown in Fig. 3-36.

**Float Traps.** Float traps are similar in appearance to F&T traps. Float traps depend entirely upon the liquid level in the trap for operation. The temperature of the condensate does not affect the operation of the trap. When sufficient water or condensate enters the trap, the float rises and lifts the valve pin off the valve seat sufficiently to allow the liquid to be discharged into the return piping. The sealing of the pin on its seat against leakage does not present a problem because the steam pressure in the trap is always higher than the pressure in the return. An external bypass, using a thermostatic trap, can be installed around a float trap, if needed. Fig. 3-37 illustrates some typical applications for F&T traps and float traps.

OPERATION

PRIMING: The inverted bucket of trap must be submerged to operate. On new installsions, trap should be primed by pouring water thru the plug hole on cover.

FIG. A

INLET   AIR BUBBLES THRU WATER   OUTLET

START-UP: Figure A shows the position of the bucket when steam is turned on. Excess condensate and air flow thru the valve seat.

FIG. B

INLET          STEAM          OUTLET

VALVE CLOSES: Figure B shows the position of the bucket when condensate ceases to flow and steam enters the trap, and accumulates in the bucket. The steam pressure will push out more condensate until enough water has been displaced by steam in the bucket and it floats, holding the valve seat closed.

FIG. C

INLET                    OUTLET

STEAM CONDENSES: Figure C steam slowly condensed inside the bucket and entrained air gradually escapes through the vent in the bucket (V), rising to the top of the trap. When steam condenses the water level rises inside the bucket causing the bucket to lose bouyancy

FIG. D

INLET                    OUTLET

VALVE OPENS: Figure D when the weight of the bucket times the lever arm ratio equals the pressure on the seat port, the bucket drops. Condensate and air are forced from the trap by steam pressure until sufficient steam has displaced the water in the bucket causing the bucket to become buoyant.

Courtesy Hoffman Specialty ITT

Fig. 3-36. An inverted-bucket trap.

RELIEF VALVE

HOT WATER

TEMPERATURE REGULATOR

THERMOMETER

DIRT STRAINER

STEAM

STORAGE TANK
WITH STEAM COIL

F & T TRAP

TO RETURN

COLD WATER

**LARGE STORAGE TANK HEATER**

SUPPLY LINE

UNIT HEATER

**UNIT HEATER**

F & T TRAP

SEDIMENT POCKET

RETURN

**DRAINING WATER FROM
COMPRESSED AIR
RECEIVER**

$\frac{3}{8}''$

$\frac{3}{4}''$ OR 1''

F & T TRAP

Fig. 3-37. Some typical applications of float

TEMPERATURE REGULATOR

THERMOMETER

STEAM

HOT WATER

DIRT STRAINER

F & T TRAP    TO RETURN    COLD WATER

**LARGE TANKLESS HEATER**

GLOBE VALVE FOR HAND
THROTTLING IN EMERGENCY    BYPASS    TEMPERATURE REGULATOR

DIRT STRAINER

GATE VALVE

HIGH-
PRESSURE
STEAM

HOT-WATER
SUPPLY MAIN

DEAD-END PRESSURE
REDUCING VALVE

FEELER PIPE    THROTTLE VALVE    F & T TRAPS    CONVERTER

**CONVERTER FOR HEATING WATER WITH HIGH PRESSURE STEAM**

AIR-SUPPLY PIPE

3/8 "

3/4 " OR 1"

F & T TRAP

TO DRAIN

**COMPRESSED AIR SUPPLY PIPE**

**Courtesy Hoffman Specialty ITT**

and thermostatic traps and float traps.

# TYPES OF TWO-PIPE SYSTEMS

Fig. 3-38 shows a typical gravity installation. A continuous-flow float and thermostatic trap is shown ahead of the meter with a float-type air vent at the trap discharge. This vent discharges all the air from the system because some meter constructions will not handle air.

Fig. 3-39 shows the installation where an intermittent type of bucket trap is used. A receiver must always be used with an intermittent operating trap so as not to overload the meter when discharging. The float-type vent is also used on the receiver.

Fig. 3-40 shows the installation of a float and thermostatic trap ahead of the meter for a vacuum installation. The bypass air vent line around the meter is required for this vacuum installation.

Courtesy Hoffman Specialty ITT

Fig. 3-38. A typical gravity installation.

Courtesy Hoffman Specialty ITT

Fig. 3-39. Installation using an intermittent bucket trap.

Fig. 3-40. Installation of a float and thermostatic trap ahead of the meter for a vacuum installation.

Fig. 3-41 shows a vacuum installation without the use of a master trap. This can be used when other traps in the system are performing without leakage. The air bypass piping is also required.

Table 3-1 gives dimension A for a variety of meter sizes for installation as shown by Figs. 3-39 through 3-41.

Fig. 3-41. A vacuum installation without the use of a master trap.

## Table 3-1. Dimension "A" for Various Meter Sizes

| METER CAPACITY LB. PER HR. | 250 | 500 | 750 | 1500 | 3000 | 6500 | 12000 |
|---|---|---|---|---|---|---|---|
| | | | DIMENSION A—INCHES | | | | |
| FIG. 39 | 4 | 4 | 8 | 8 | 8 | 12 | 12 |
| FIG. 40 | 12 | 12 | 15 | 15 | 15 | 18 | 18 |
| FIG. 41 | 12 | 12 | 15 | 15 | 15 | 18 | 18 |

113

Local codes in most cities do not permit the discharge of condensate into the sewer at temperatures above a specific amount. This is usually between 140°F and 150°F. In order to assure control of this maximum temperature requirement and to extract all the heat possible from the condensate before discharging it to the sewer, a method of salvaging this heat is required.

Fig. 3-42 shows a method of discharging the condensate through an economizer to preheat domestic water before it goes to the primary water heater. If the condensate is pumped, it should first be discharged into a vented receiver from which it can flow by gravity through the economizer, or preheater, and the flowmeter.

Courtesy Hoffman Specialty ITT

Fig. 3-42. Discharging the condensate through an economizer to preheat domestic water before it goes to the primary water heater.

## Gravity Condensate Return System

The two basic two-pipe steam heating systems are the gravity return system and the mechanical return system. The diagram of a simple gravity return system is shown in Fig. 3-43. The definition and function of the system components are explained referring to Fig. 3-43.

SUPPLY VALVE

THERMOSTATIC TRAP

(B) MAIN SUPPLY

(A) HEADER

MAIN VENT

A

F & T TRAP

WATER LINE

(C) DRY RETURN

2"

BOILER

(D) WET RETURN

(E) HARTFORD LOOP    CLOSE NIPPLE

**Fig. 3-43. The main components of a two-pipe gravity return system.**

(A) The steam-supply header is the horizontal piping connecting to one or more steam outlets at the top of the boiler. The
steam-supply main or mains connect to the header.

(B) The steam-supply main carries the steam from the boiler
and header to the radiators or other heating units in the
system. As shown in Fig. 3-43 an up-feed runout carries
steam to the radiator through a supply valve connected to
the inlet connection. A thermostatic trap is connected to
the outlet connection on the radiator. The return runout
piping carries the trap discharge to the dry-return main.

(C) Part of the return main is called a *dry return* to distinguish it
from the *wet return*. As shown in Fig. 3-43, the *dry return* is
located above the boiler water line. The dry return carries
the air and condensate discharge from the radiator thermostatic trap and from the F & T trap, which drips the end
of the steam-supply main.

(D) The *wet return* is the portion of the return main that carries
condensate back to the boiler and is installed below the
boiler water line. Again referring to Fig. 3-43, it will be seen
that the wet return is completely filled with water.

115

(E) The purpose of the Hartford loop is to prevent the water from leaving the boiler through the return piping. The construction details of a Hartford loop are explained in the one-pipe section of this chapter.

To sum up the basic differences between a one-pipe gravity return system and a two-pipe gravity return system, the two-pipe system requires a separate return line. Because the supply main is used as a steam-supply line *only* in a two-pipe system, a thermostatic or a float and thermostatic trap must be installed on the end of the steam-supply main.

## Mechanical Condensate Return System

As steam heating systems increase in size, requiring higher steam pressures in order to obtain steam circulation, some means other than gravity must be used to return the steam condensate to the boiler. The use of a condensate pump for this purpose has become the accepted method. The condensate from the return piping system flows by gravity into a tank or receiver. A condensate pump (or pumps) mounted on or adjacent to the receiver pumps the accumulated condensate into the boiler return piping. The return main must be graded uniformly to a slope of ¼" in 10' when carrying the condensate to the receiver to ensure that con-

AIR VENT

FROM RETURNS

CHECK VALVE

TO BOILER

PUMP

RECEIVER

Courtesy Hoffman Specialty ITT

Fig. 3-44. A condensate receiver filling with water.

densate and air will be separated in the piping. The air can then be vented into the atmosphere through the receiver vent connection. If it is necessary because of a job condition or to obtain uniform pitch to the receiver inlet, the condensate receiver can be located underground or in a pit. Fig. 3-44 shows a receiver filling with condensate. As the water level rises, the float also rises until the point is reached where the float actuates the float switch. The float switch energizes the pump, and the water level drops as the condensate is pumped into the boiler return piping (Fig. 3-45). When the float, dropping with the water level, reaches its low position, the pump switch is turned off, and the receiver starts to accumulate condensate, beginning the cycle over again.

Fig. 3-45. Water level in a receiver dropping as condensate is pumped in the boiler return piping.

## VACUUM SYSTEMS

A vacuum system operates with a steam pressure that is less than atmospheric pressure. The object of such a system is to take advantage of the low working temperature of the steam at this low pressure, giving a mild form of heat similar to that obtained with hot-water heating systems.

117

## Vacuum Return-Line System

A vacuum return-line system is similar to a condensate return system except that a vacuum pump is installed to provide a low vacuum in the return line to return the condensate to the boiler. This is shown in Fig. 3-46. Pressure in the steam-supply lines and in the radiators is maintained in a positive manner. This system is adaptable to any size or type of building.

Some of the advantages of a vacuum return-line steam heating system are:

1. A positive return of the condensate to the boiler.
2. Smaller pipe sizes can be used because of the greater pressure differential between the supply and return lines.
3. The air is removed from the steam mechanically, resulting in a rapid circulation of the steam.

One of the major disadvantages of this type of steam is the relatively high cost of the vacuum pump, which may be more than the average person cares to invest.

## Vacuum Air-Line System

The vacuum air-line system, shown in Fig. 3-47, is a variation of the well-known one-pipe steam system. The radiator air vents are replaced with air valves, the outlets of which are connected to a return air line. A vacuum pump is included to exhaust the system of air. Air-line valves are of the thermostatic type.

The advantages of this type of steam heating system are:

1. The radiators heat efficiently at lower pressures.
2. Steam circulation in the system is more rapid.
3. Air vents, which often have an objectionable noise factor, are not required.

The disadvantages of this type of system are:

1. The steam is noisy at times, primarily because the steam and condensate flow in the same pipe.
2. Radiator valves and piping must be oversize to accommodate the flow of both steam and condensate.

Courtesy Dunham-Bush, Inc.

Fig. 3-46. Piping of a vacuum-return-line steam heating system.

**Fig. 3-47.** An air-line pump installed on a one-pipe steam system to create a vacuum air-line system.

**Fig. 3-48. A typical variable-vacuum steam heating system installation.**

## Variable Vacuum System

A variable vacuum system (Fig. 3-48) can better regulate and vary the steam pressure in the radiators, thereby controlling the radiator temperature. The radiator temperature can be regulated so as to be similar to hot-water radiator temperatures. Steam pressure can range from a high vacuum (low absolute pressure) to a positive value above atmospheric pressure. A continuously controlled steam supply can be maintained by regulating its generation or by regulating its admission to the steam mains. The vacuum-pump controller regulates and maintains a slight differential between the supply piping and the return piping—the return is always slightly lower to speed the steam circulation when operating at positive pressures.

If lift connections are used, they are restricted to a single lift between the accumulator tank and the vacuum pump. It is recommended that the piping be installed so that all condensate flows by gravity to the accumulator tank.

The principal advantages of a variable vacuum steam heating system are:

1. A uniform floor-to-ceiling temperature can be maintained.
2. The types of radiation can be varied with maximum efficiency expected.
3. The system can be operated at much higher temperatures resulting in more efficiency during extreme cold weather.
4. Heat distribution can be balanced.
5. The system responds rapidly to temperature changes.
6. The steam supply is continuous, creating a better indoor climate without extreme temperature fluctuations.

The main disadvantage of this system is the relatively high cost of the vacuum pump for small heating systems. This may not make it competitive in price with other installations.

CHAPTER 4

# Steam Heating System Design

Steam heating systems are identified by one or more of the following combinations, which relate to the features of the system: (1) piping arrangement, (2) pressure range, and (3) methods employed for the return of condensate.

1. When identified by piping arrangement, they can be either one-pipe systems or two-pipe systems. They can be either up-feed or down-feed depending on the direction of steam flow in the risers. They can be described as a wet-return system or a dry-return system depending on the location of the condensate return main above or below the water line of the boiler or of the condensate receiver.

2. Systems are called vapor systems when they operate at pressures ranging from low pressure to vacuum without the use of a vacuum pump. It is a vacuum system when it operates over

123

this same pressure range using a vacuum pump. A system is a low-pressure system when it employs pressure ranging from 0 to 15 psig. It is a high-pressure system when it operates at pressure above 15 psig.

3. Systems are described as gravity systems when condensate returns directly to the boiler by gravity against the pressure causing the steam flow. When the condensate cannot be returned by gravity, mechanical means must be used to return the condensate accumulation to the boiler. The return mains can be at atmospheric pressure for nonvacuum systems or under vacuum when a vacuum pump is used. In either system, using mechanical means for condensate return, the condensate must flow by gravity to the receiver.

## DESIGN PRESSURE DROPS

Steam distribution in a heating system depends on the pressure drop selected for the system. The steam-supply piping should be sized so that the pressure drop for the developed length of branches from the same supply are as nearly uniform as possible for the same supply pressure. The pressure drop is expressed in a unit length of 100 ′ of equivalent length of run.

The pressure drop in common use for steam piping is as follows:

| SYSTEM PRESSURE psig | PRESSURE DROP per 100 ft. | TOTAL SYSTEM PRESSURE DROP IN STEAM SUPPLY PIPING |
|---|---|---|
| Vacuum Systems | 2–4 oz. | 1–2 psi |
| 1 | 2 oz. | 1–4 oz. |
| 2 | 2 oz. | 8 oz. |
| 5 | 4 oz. | 1½ psi |
| 10 | 8 oz. | 3 psi |
| 15 | 1 psi | 4 psi |

Return piping is always sized for the same pressure drop as selected for the supply piping.

Arranging the system piping so that the total distance of the supply from the boiler to the radiator is the same as the distance of the return from the radiator back to the boiler gives the desirable resistance to and from the radiator. It is important that the piping be sized to handle the full design-load conditions. A heating system operates at less than half of its design-load conditions during an average winter. However, the pickup load required to raise the temperature of the metal in the piping up to the steam temperature and to raise the indoor temperature up to design conditions requires a large amount of heat even during moderate winter outdoor temperature. This increased pickup value is usually considered to be 143% of the design load.

## SIZING BOILER CONNECTIONS AND HEADER PIPING

Steam boilers for heating systems have one or more outlets. When there is more than one, all the outlets should be used. The vertical pipe to the supply main or the header should be the same size as the boiler outlet tapping and never reduced except at the supply main or header. The header should be sized on the basis of the maximum load that must be carried by any part of it. The sizing can be done from the same capacity table as used for sizing the steam mains. Increasing the size of the header to a larger pipe size than required can be undesirable when the pickup load during the heating-up period is considered.

The return header is the same size as the required return main at the boiler. If the boiler has more than one return tapping, both should be used and sized the same as the return main.

## DESIGNING A ONE-PIPE STEAM SYSTEM

The piping for a one-pipe gravity air vent system in which the total equivalent length of run does not exceed 200 ' can be sized from Tables 4-3, 4-5, and 4-7 at the end of this chapter. Instructions for their use is as follows:

1. Where steam and condensate flow in the same direction in the steam main, dripped runouts to up-feed risers, and down-

125

feed risers, use Table 4-3, Col. 2 (1 oz. per 100′ pressure drop).

2. Where runouts to risers are not dripped and steam and condensate flow in the opposite directions, use Table 4-5, Col. F. Use the same for runouts to radiators.
3. For up-feed risers carrying condensate back from the radiators, use Table 4-5, Col. D.
4. For down-feed systems, the main riser between the boiler and the overhead attic main, which does not carry condensate from the radiators, use Table 4-5, Col. B.
5. Size radiators supply valves and vertical connections from Table 4-5, Col. E.
6. Use Table 4-7, Col. O, for sizing dry-return mains.
7. Use Table 4-7, Col. N, for sizing wet-return mains.

Additional notes on the installation of the piping for a one-pipe system are as follows.

1. Supply and dry-return mains should not be pitched less than 4″ in 10′.
2. Horizontal runouts to risers and radiators should not be less than 4″ per foot. Where this pitch cannot be obtained for runouts over 8′ in length, increase the pipe one size.
3. It is not desirable to have a supply main smaller than 2″.
4. All supply mains, runouts to risers, or risers should be dripped where necessary.
5. When supply mains are decreased in size, they should be dripped or eccentric reducer coupling should be used as shown by Fig. 4-19B.

The following 6 steps are required for a complete design of a one-pipe steam system.

1. Calculate the heat loss for each room or space. Use *ASHRAE Guide, 1 = B = R*, or other available publications. Heat-loss calculations are not performed for this example.
2. Decide on type and size of heating units for each room. Consult *ASHRAE Guide* and manufacturers' catalogs for specific information.
   A. Convert the Btu/hr heat loss for each room to Equivalent Direct Radiation—1 EDR = 240 Btu/hr.

B. Locate heating unit to suit space and wall exposure. This location is usually on exposed outside walls and under or near windows so as to replace heat losses where they occur.

NOTE: *Example shows size only of required heat output for a unit and not a specific make of unit.*

3. Make a piping layout to obtain the most direct piping arrangement for supplying steam from the boiler to the most remote radiator, and for the return of condensate. Usually a double-circuit main has the advantage of quick and uniform delivery of steam to the heating units. Fig. 4-1 shows the piping layout for the two-circuit system.

A. Measure actual length of main and risers from the boiler to most remote radiator to establish maximum pipe length:

$$\begin{aligned}
\text{Section ABCG} &= 103' \\
\text{Riser (1)} &= 20' \\
\text{Total} &= 123'
\end{aligned}$$

B. Determine total equivalent length of piping by making proper allowance for fittings—see Table 4-6.

Main
2—2½" elbows                = 10'
3—2½" side outlet tees   = 33'
2—2" elbows                   = 8.6'
Riser
1—2" side outlet tee       = 8'
1—½" side outlet tee      = 7'
3—½" elbows                 = 10.5'
                                 Total = 77.1'
Total Equivalent Length = 123' + 77' = 200'

4. Determine system pressure drop and boiler pressure:

A. For gravity system where equivalent length is less than 200' the total pressure drop should not exceed 2 oz. For equivalent length greater than 200' the total pressure drop should not exceed 4 oz. Because the total equivalent length of this example is not more than 200′, the piping will be sized on the basis of a total pressure drop of 2 oz.

B. Boiler pressure should not be less than twice the total pressure drop. The boiler pressure will be maintained at a minimum practical control limit of ¼ psi.

**Fig. 4-1. A piping layout for a two-circuit system.**

5. Size supply mains, risers, return piping.

A. Divide total system pressure drop by equivalent length of run to obtain pressure drop for 100′ length.

B. Consult Table 4-3, Col. 2, for supply main sizes. The minimum horizontal supply main should not be less than 2″.

C. Consult Table 4-7, Col. N, for wet-return main sizes. Size returns on basis of same pressure drop for 100′ length as that selected for supply main.

D. Consult Table 4-5 for sizing the following pipe connections:

(1) Use Col. F for horizontal runout to up-feed riser or to radiator.

(2) Use Col. E for vertical connection to radiator and radiator supply valve size.

(3) Use Col. D for up-feed supply riser.

6. Select a boiler from manufacturer's catalog having a net I = B = R rating equal to or greater than the total connected load; in this example 535 EDR. If domestic hot water is to be heated from this steam boiler, no allowance need be made unless there are more than two bathrooms. Consult I = B = R recommendation for additional allowance when required.

## Pipe Sizes for One-Pipe System

| DESCRIPTION | SECTION | LOAD—<br>Sq. Ft. EDR | PIPE<br>SIZE |
|---|---|---|---|
| Horizontal Runout to Radiator | | 35 | 1¼" |
| Horizontal Runout to Radiator | | 40 | 1¼" |
| Horizontal Runout to Radiator | | 50 | 1¼" |
| Vertical Connection Radiator Supply Valve | | 35 | 1¼" |
| Vertical Connection Radiator Supply Valve | | 40 | 1¼" |
| Vertical Connection Radiator Supply Valve | | 50 | 1¼" |
| Up-feed Riser | (1) | 100 | 1½" |
| Up-feed Riser | (4) | 80 | 1¼" |
| Up-feed Riser | (2) & (3) | 50 | 1¼" |
| Supply Main | C to R | 280 | 2" |
| Supply Main | C to S | 255 | 2" |
| Supply Main | B to C | 535 | 2½" |
| Boiler Header | A to B | 535 | 2½" |
| Wet-Return Main | R to T | 280 | 1" |
| Wet-Return Main | S to T | 235 | 1" |

## DESIGNING A TWO-PIPE SYSTEM

Two-pipe steam heating systems can be designed with piping arranged for an up-feed system or for a down-feed system from a steam-supply main located in an attic or pipe space. All two-pipe systems are similar in all basic respects. Attention must be given to the details of dripping the ends of the steam main, the bottom or heel of up-feed risers through steam drip traps.

The use of a condensate pump for gravity open-vented return system or the use of vacuum pumps for a vacuum system is the preferred method of returning condensate to the boiler. Using these mechanical means eliminates the problem for the system. Capacity tables for sizing steam and return piping are contained in the section on Engineering Data and Technical Information.

Additional information for the piping arrangement for two-pipe systems is as follows:

1. Steam and return mains should be pitched not less than ¼″ in 10′.
2. Horizontal runouts to risers and radiators should be pitched not less than ½″ per foot. Where this pitch cannot be obtained for runouts over 8′ in length, increase the pipe one size larger than called for in the capacity table.
3. It is not desirable to have a steam-supply main that is smaller than 2″.
4. When required, the ends of steam mains, bottom of up-feed riser or runouts to up-feed risers should be dripped through steam traps into the dry return main. The ends of all down-feed steam risers should also be dripped through steam traps to the dry return.
5. Return mains should always be pitched to provide gravity flow of condensate to the pump receiver inlet.

The following eight steps are required for a complete design of a two-pipe steam heating system. The example shown in Fig. 4-2 will be used to illustrate the step-by-step procedure. Figs. 4-3, 4-4, and 4-5 are enlarged sections of Fig. 4-2.

1. Calculate the heat loss for each room or space. Many publications are available which can be used to determine the heat loss, such as *ASHRAE Guide and Data Book*, and I = B = R *Heat Loss Calculation Guide*. Heat-loss calculations are not performed for this example.
2. Decide on the type of heating unit to be used and the size required for each room. Specific information must be obtained from manufacturers' literature and catalog ratings.

With this information available, it can be used as follows:

A. Convert the Btu/hr heat loss from each room to square feet Equivalent Direct Radiation. 1 square foot EDR = 240 Btu/hr.

B. Locate the heating units on the floor plan to suit available space and wall exposure. The preferable location is on an outside wall and under or near windows. This is done so as to replace the heat losses where they occur and to offset them where they have the highest rate of loss. The example shows the size of the heating units in square feet EDR but does not specify the type of unit.

3. Determine equivalent length of piping.

A. Make a piping layout to obtain the most direct piping arrangement for supplying steam from the boiler to the most remote radiator and for the return of condensate to the boiler. For this example, the application of a heating system to a long, narrow, two-story building was chosen, as shown by Fig. 4-2, with the boiler located in one end of the building. The supply and return piping for the radiators are connected to obtain the shortest length of pipe for steam flow from the radiators. The steam-supply piping begins at the boiler and ends at the most remote radiator, while the return piping begins at the most remote radiator and ends at the condensate pump adjacent to the boiler. When the piping arrangement has been determined, the procedure is as follows:

(1) Measure the actual length of steam-supply main, runouts, and risers from the boiler to the most remote radiator to establish the maximum pipe length, as shown by Fig. 4-2.

(2) Determine the total equivalent length of piping by making proper allowance for fittings. Table 4-6 shows the friction allowance for fittings that must be added to the measured length to determine the equivalent length. Because the selection of pipe sizes for the system is based on the total equivalent length of piping, it is necessary to begin the design process based on an assumed equivalent length. This assumed

131

## Tabulation of Return-Pipe Sizes for Fig. 4-2

| PIPE SECTION LOCATION | SQUARE FEET EDR | | PIPE SIZE (inches) | REMARKS |
|---|---|---|---|---|
| | ACTUAL LOAD | PERMITTED LOAD | | |
| D6 to D2 | 455 | 460 | 1 | |
| D1 to C2 | 900 | 964 | 1¼ | |
| C1 to B1 | 1450 | 1512 | 1½ | |
| B1 to Pump Receiver | 1450 | 1512 | 1½ | Table 4-7—Col. U |
| Runout to Riser* | 115 | 460 | 1 (min.) | |
| Riser* | 115 | 460 | 1 | |
| Runout to Riser† | 150 | 460 | 1 | |
| Riser† | 150 | 460 | 1 | |

 * See Fig. 4-3.
 † See Fig. 4-4.

## Tabulation of Steam-Pipe Sizes for Fig. 4-2

| PIPE SECTION LOCATION | SQUARE FEET EDR | | PIPE SIZE (inches) | REMARKS |
|---|---|---|---|---|
| | ACTUAL LOAD | PERMITTED LOAD | | |
| D6 to D1 | 520 | 936 | 2 | Table 4-3—Col. 6 |
| C4 to C2 | 900 | 936 | 2 | Table 4-3—Col. 6 |
| C1 to B1 | 1450 | 1512 | 2½ | Table 4-3—Col. 6 |
| Boiler Header | 1450 | 1512 | 2½ | Table 4-3—Col. 6 |
| Runout to Riser* | 115 | 152.4 | 1¼ | Table 4-4—5" Pitch |
| Riser* | 115 | 148 | 1 | Table 4-3—Col. 6 |
| Runout to Riser† | 150 | 152.4 | 1¼ | Table 4-4—5" Pitch |
| Riser† | 150 | 312 | 1¼ | Table 4-3—Col. 6 |

 * See Fig. 4-3.
 † See Fig. 4-4.

## Tabulation of Radiator Connection for Fig. 4-2

| SQUARE FEET EDR LOAD | STEAM SUPPLY* | | | RETURN PIPE† | | |
|---|---|---|---|---|---|---|
| | Runout from Riser or Main (inches) | Vertical Connection (inches) | Valve Size (inches) | Runout from Riser or Main (inches) | Vertical Connection (inches) | Trap Size (inches) |
| 50 | 1 | ¾ | ¾ | ¾ | ½ | ½ |
| 65 | 1 | ¾ | ¾ | ¾ | ½ | ½ |
| 80 | 1 | ¾ | ¾ | ¾ | ½ | ½ |

 *From Table 4-8.
 **From Table 4-9.
 NOTE: Fig. 4-5 shows detailed pipe sizing at end of main and riser.

Fig. 4-2. A complete design of a two-pipe steam heating system.

**Fig. 4-3. Sections 1 and 2 of the plan shown in Fig.4-2.**

value is usually considered to be double the actual measured length. For this example, the assumed equivalent length will be 250' × 2 = 500'. The actual equivalent length of pipe can be checked after the piping has been sized, based on the assumed length.

4. Determine system pressure. Before pipe sizes can be selected for the system, several factors must be considered that pertain to the steam pressure at the boiler and system pressure drop under full-load conditions. These factors include the following:

    A. The total system pressure drop must not exceed ½ the boiler pressure when steam and condensate flow in the same direction.

    B. The pressure drop must be kept at a value that will not create excessive velocities. For systems operating below

15 psi, the velocities should not be more than 5000 ' per minute for quiet operation for concurrent flow. For the example shown in Fig. 4-2, the boiler pressure of 3 psi was selected for full-load conditions. The system pressure drop, therefore, must not exceed 1.5 psi (24 oz.).

5. Size the supply and return piping.

   A. From step 3 the equivalent length of piping was assumed to be 500'. From step 4 the maximum system pressure drop has been established as 1.5 psi (24 oz.). Therefore, for 500' equivalent length, the pressure drop per 100 ' of length will be approximately 5 oz. However, Table 4-3, Col. 6, shows the nearest pressure to be 4 oz. per 100 ' and

Courtesy Hoffman Specialty ITT

**Fig. 4-4. Sections 3 and 4 of the plan shown in Fig. 4-2.**

135

Courtesy Hoffman Specialty ITT

**Fig. 4-5. Sections 5 and 6 of the plan shown in Fig. 4-2.**

the capacities shown in this column will be used to size the steam main. When sizing supply mains for low-pressure steam heating systems, it is not desirable to have a horizontal supply main smaller than 2″.

B. The horizontal steam runout to risers and risers will be sized from Table 4-3, Col. 6, and Table 4-4, 5″ pitch column.

C. The return mains, horizontal runouts to risers, and the return risers will be sized from Table 4-7, Col. U for a pressure drop of 2 psi (4 oz.), which is the same as that used for sizing the steam piping.

D. The horizontal steam runout to radiators from steam

main or riser and the supply valve vertical connection and valve size are selected from Table 4-8. The horizontal return runouts to radiators from return mains or risers and radiator trap sizes are selected from Table 4-9.

E. Referring to Fig. 4-2 and using the pipe capacity tables, the following tabulation shows the pipe sizes selected for the design example. The sizes are also shown in the piping layout of Fig. 4-2.

F. Pipe sizing up to this point has been made on the basis of an assumed equivalent length. The actual equivalent length can now be determined and the pressure drop per 100′ of pipe length can also be determined. The following tabulation shows the result of this check for system pressure drop.

Steam main measured length 250′.

### Allowance for Fittings

Steam Main

| | |
|---|---:|
| 1—2½″ Ell | 5.0 |
| 2—2½″ S.O. Tee | 22.0 |
| 1—2½″ G.V. | 1.1 |
| 3—2½″ Ell | 15.0 |
| 1—2½″ S.O. Tee | 11.0 |
| 1—2″ Ell | 4.3 |
| 2—2″ S.O. Tee | 16.0 |
| Runout from Main | |
| 2—1″ Ell | 4.4 |
| Runout from Riser | |
| 1—1″ S.O. Tee | 5.0 |
| 2—1″ Ell | 4.4 |
| 1—¾″ Angle Valve | 10.0 |
| | 348.2′ |

As can be seen, with an actual total equivalent length of 348.2′ and a total system pressure drop of 24 oz. (1.5 psi), the pressure drop per 100 feet is 6.8 oz. Therefore, a study of Table 4-3 shows that there will be no changes in pipe size if the sizing is made on the basis of an 8-oz. drop (Col. 8) instead of a 4-oz. drop (Col. 6).

137

G. Determine steam velocity in the system with 3 psi boiler pressure, 24″ main pipe size, 362 lb./hr (lb./hr. = EDR/4) load in system. 5000′ per minute maximum. By calculation using steam velocity formula in Engineering Data and Technical Information Section we get a velocity of 3996′ per minute.

6. Select a boiler from manufacturer's catalog having a net rating (I = B = R or SBI) equal to or greater than total connected load to 1450 square feet EDR. Often, domestic water is heated by the same boiler used to supply steam for heating. The size of the heater and storage tank depends on the water requirements. Proper sizing information is available from many sources such as *ASHRAE Guide*, I = B = R, SBI, and several manufacturers of heaters and coils.

7. Select float and thermostatic traps to drip the main.

> MAIN A-C = 470 sq. ft.
> MAIN C-D = 460 sq. ft.
> MAIN D-E = 520 sq. ft.

A Hoffman ¾″ No. 53FT trap will handle 560 square feet EDR(140 lb. of condensate per hour) with a pressure differential of 1 psi.

8. Select a condensate pump from Hoffman Condensate catalog to meet system requirements.

> Heating system load      1450 sq. ft
> Boiler pressure      3 psig

# PIPING ILLUSTRATIONS FOR ONE-PIPE AND TWO-PIPE STEAM SYSTEMS

Figs. 4-6 through 4-21, which follow, are illustrations made to supply additional information required for the design of one- and two-pipe steam heating systems.

(A) Correct method of making up-feed connection to radiator from one-pipe steam system.

(B) Correct method of connecting radiator to up-feed or down-feed riser from one-pipe steam system.

Courtesy Hoffman Specialty ITT

**Fig. 4-6. Radiator connections for one-pipe steam systems.**

## ENGINEERING DATA AND TECHNICAL INFORMATION

This section contains necessary engineering technical information used in the design, installation, maintenance, or revamping of low-pressure steam heating systems. The tables, charts, and examples include properties of saturated steam, steam capacities of ASTM Schedule 40 (S) pipe and return pipe capacities used in one-pipe and two-pipe steam heating systems. Also included is data for determining pipe friction allowance, steam velocity, system pressure drops, and similar information.

### Velocity of Steam

To find the approximate velocity of low-pressure steam (feet per second) through a pipe, multiply the condensation in pounds per hour by the volume of steam in cubic feet per pound corresponding to the steam pressure. Divide this result by 25 times the internal area of the pipe.

The pipe area is found in the table "Standard Pipe Dimensions" and the volume per cubic foot is found in the table "Properties of Saturated Steam."

139

SUPPLY MAIN

VENT VALVE

PACKLESS RADIATOR VALVE

REDUCER

NOT LESS THAN 18″

WATER LINE OF BOILER

CAN BE REDUCED (NEVER LESS THAN ¾″)    WET RETURN

DOWN-FEED CONNECTIONS WITH WET RETURN

(A) Correct method of making down-feed connection to radiator and to wet return.

PACKLESS RADIATOR VALVE    TRAP

MINIMUM COOLING LEG 5′ 0″ LONG    DRY RETURN

DOWN-FEED CONNECTIONS DRIPPING RISER INTO CONDENSATION PUMP RETURN

(B) Correct method of making down-feed connection to radiator and to dry return.

Courtesy Hoffman Specialty ITT

Fig. 4-7. Radiator connections for one-pipe systems.

STRAIGHT-SHANK AIR VALVE

AIR VALVE
AIR CHAMBER

PITCH DOWN ⇨

AIR VALVE
AIR CHAMBER    SUPPLY VALVE OR UNION

PACKLESS RADIATOR VALVE

VENTING CONCEALED RADIATOR OF HEADER TYPE—AIR VENT TAPPING IN TOP

PITCH DOWN ⇨

SUPPLY VALVE OR UNION

Courtesy Hoffman Specialty ITT

Fig. 4-8. Three methods of reducing radiation to suit varying job conditions (one-pipe systems).

(A) The proper method of making up-feed connections to a radiator.

(B) The proper method of making radiator connections taken from up-feed or down-feed risers.

**Courtesy Hoffman Specialty ITT**

**Fig. 4-9. Radiator connections for two-pipe systems.**

**Courtesy Hoffman Specialty ITT**

**Fig. 4-10. The proper method for connecting a radiator to an overhead supply main (two-pipe system).**

141

(A) Unit heater connections for a two-pipe gravity or vacuum system.

(B) Unit heater connections for a two-pipe gravity or vacuum system with the supply branch dripped through a trap.

Courtesy Hoffman Specialty ITT

Fig. 4-11. Unit heater connections for a two-pipe system.

(A) The proper method of connecting a radiator to a down-feed riser, dripping the heel of the riser into the dry return.

(B) The proper method of connecting a radiator to concealed risers.

Courtesy Hoffman Specialty ITT

Fig. 4-12. Radiator connections for a two-pipe system.

(A) The proper method of connecting a ceiling radiator located above the supply and return piping.

(B) The proper method of connecting a ceiling radiator with the return bled into the wet return.

Courtesy Hoffman Specialty ITT

**Fig. 4-13. Radiator connections for a two-pipe steam system.**

Courtesy Hoffman Specialty ITT

**Fig. 4-14. Various methods of connecting convertors to supply-and-return piping (two-pipe systems).**

143

Fig. 4-15. The correct method of connecting exposed steam coils to supply-and-return piping (two-pipe systems).

Fig. 4-16. The correct method of connecting blast coils to thermostatic supply valves to return-and-supply piping.

144

(A) Correct method of dripping end of supply main into the dry return main.

(B) Correct method of dripping drop riser or end of main into return.

(C) Correct method to connect up-feed risers.

(D) Correct method for reducing size of main.

Courtesy Hoffman Specialty ITT

**Fig. 4-17. Trap installations for two-pipe systems.**

LOW-PRESSURE CLOSED GRAVITY SYSTEM

HIGH-PRESSURE SYSTEM

VACUUM OR LOW-PRESSURE OPEN GRAVITY SYSTEM

Courtesy Hoffman Specialty ITT

**Fig. 4-18. The correct method of connecting unit heaters to low-pressure closed systems, low-pressure open systems, and high-pressure systems.**

145

ACCEPTABLE    PREFERRED

ECCENTRIC REDUCER

(A) The acceptable and the preferred methods of taking branches from mains.

(B) The correct method, using an eccentric reducer, of reducing the size of mains.

(C) The correct method of looping a main around a beam.

(D) The correct method of looping a dry-return main around a door.

| B | CONSTANT |
|---|----------|
| 11-¼° | 5.126 |
| 22-½° | 2.613 |
| 30° | 2.000 |
| 45° | 1.414 |
| 60° | 1.155 |

TO FIND C, MULTIPLY A BY CONSTANT FOR ANGLE B

(E) How to use constants to determine length of pipe in break when making offsets in piping.

Courtesy Hoffman Specialty ITT

Fig. 4-19. The correct methods of solving common piping problems.

(A) An up-feed branch connection taken from main at a 45° angle.

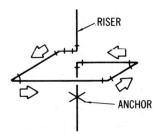

(B) An expansion joint made up of pipe for risers.

(C) A drop riser branch taken from the top of a main at a 45° angle.

(D) A drop riser branch taken from the bottom of a main.

(E) A riser branch taken from the bottom of a main and dripped into the wet return.

(F) The correct method of taking double radiator branch connections from a riser.

Courtesy Hoffman Specialty ITT

**Fig. 4-20. Ways of solving common piping problems.**

(A) Correct method of dripping the end of a one-pipe steam main which extends past the end of the wet return.

(B) Correct method of making expansion joints of piping.

Fig. 4-21. Ways of solving common piping problems.

*Example.* What is the velocity of steam at 5 lbs. per sq. inch flowing through a 2″ pipe at a rate to produce 175 lbs. of condensate per hour?

1 pound of steam at 5 lbs pressure = 20 cu. ft.

Internal area of 2″ pipe = 3.36

The velocity of steam is:

$$\frac{175 \times 20}{25 \times 3.36} = \frac{3500}{84} = 41.7 \text{ ft. per sec.}$$

Table 4-3 gives the capacity of ASTM Schedule 40 (S) pipe expressed in square feet EDR. The values are obtained from charts published by the American Society of Heating, Refrigerating and Air Conditioning Engineers' *1967 Guide.*

Where condensate must flow counter to steam flow, the governing factor is the velocity that will not interfere with the condensate

### Table 4-1. Properties of Saturated Steam (Approx.)

| Absolute Pressure | Gage Reading at Sea Level | Temp. °F. | Heat in Water Btu per Lb. | Latent Heat In Steam (Vaporization) Btu per Lb. | Volume of 1 Lb. Steam Cu. Ft. | Wgt. of Water Lbs. per Cu. Ft. |
|---|---|---|---|---|---|---|
| 0.18 | 29.7 | 32 | 0.0 | 1076 | 3306 | 62.4 |
| 0.50 | 29.4 | 59 | 27.0 | 1061 | 1248 | 62.3 |
| 1.0 | 28.9 | 79 | 47.0 | 1049 | 653 | 62.2 |
| 2.0 | 28 | 101 | 69 | 1037 | 341 | 62.0 |
| 4.0 | 26 | 125 | 93 | 1023 | 179 | 61.7 |
| 6.0 | 24 | 141 | 109 | 1014 | 120 | 61.4 |
| 8.0 | 22 | 152 | 120 | 1007 | 93 | 61.1 |
| 10.0 | 20 | 161 | 129 | 1002 | 75 | 60.9 |
| 12.0 | 18 | 169 | 137 | 997 | 63 | 60.8 |
| 14.0 | 16 | 176 | 144 | 993 | 55 | 60.6 |
| 16.0 | 14 | 182 | 150 | 989 | 48 | 60.5 |
| 18.0 | 12 | 187 | 155 | 986 | 43 | 60.4 |
| 20.0 | 10 | 192 | 160 | 983 | 39 | 60.3 |
| 22.0 | 8 | 197 | 165 | 980 | 36 | 60.2 |
| 24.0 | 6 | 201 | 169 | 977 | 33 | 60.1 |
| 26.0 | 4 | 205 | 173 | 975 | 31 | 60.0 |
| 28.0 | 2 | 209 | 177 | 972 | 29 | 59.9 |
| 29.0 | 1 | 210 | 178 | 971 | 28 | 59.9 |
| 30.0 | 0 | 212 | 180 | 970 | 27 | 59.8 |

*Left axis label: INCHES OF MERCURY. Gage reading axis label: VACUUM—INCHES OF MERCURY.*

Courtesy Hoffman Specialty ITT

## Table 4-1. Properties of Saturated Steam (Approx.) (Cont'd)

| Absolute Pressure | Gage Reading at Sea Level | Temp. °F. | Heat in Water Btu per Lb. | Latent Heat in Steam (Vaporiza-tion) Btu per Lb. | Volume of 1 Lb. Steam Cu. Ft. | Wgt. of Water Lbs. per Cu. Ft. |
|---|---|---|---|---|---|---|
| 14.7 | 0 | 212 | 180 | 970 | 27 | 59.8 |
| 15.7 | 1 | 216 | 184 | 968 | 25 | 59.8 |
| 16.7 | 2 | 219 | 187 | 966 | 24 | 59.7 |
| 17.7 | 3 | 222 | 190 | 964 | 22 | 59.6 |
| 18.7 | 4 | 225 | 193 | 962 | 21 | 59.5 |
| 19.7 | 5 | 227 | 195 | 960 | 20 | 59.4 |
| 20.7 | 6 | 230 | 198 | 958 | 19 | 59.4 |
| 21.7 | 7 | 232 | 200 | 957 | 19 | 59.3 |
| 22.7 | 8 | 235 | 203 | 955 | 18 | 59.2 |
| 23.7 | 9 | 237 | 205 | 954 | 17 | 59.2 |
| 25 | 10 | 240 | 208 | 952 | 16 | 59.2 |
| 30 | 15 | 250 | 219 | 945 | 14 | 58.8 |
| 35 | 20 | 259 | 228 | 939 | 12 | 58.5 |
| 40 | 25 | 267 | 236 | 934 | 10 | 58.3 |
| 45 | 30 | 274 | 243 | 929 | 9 | 58.1 |
| 50 | 35 | 281 | 250 | 924 | 8 | 57.9 |
| 55 | 40 | 287 | 256 | 920 | 8 | 57.7 |
| 60 | 45 | 293 | 262 | 915 | 7 | 57.5 |
| 65 | 50 | 298 | 268 | 912 | 7 | 57.4 |
| 70 | 55 | 303 | 273 | 908 | 6 | 57.2 |
| 75 | 60 | 308 | 277 | 905 | 6 | 57.0 |
| 85 | 70 | 316 | 286 | 898 | 5 | 56.8 |
| 95 | 80 | 324 | 294 | 892 | 5 | 56.5 |
| 105 | 90 | 332 | 302 | 886 | 4 | 56.3 |
| 115 | 100 | 338 | 309 | 881 | 4 | 56.0 |
| 140 | 125 | 353 | 325 | 868 | 3 | 55.5 |

Courtesy Hoffman Specialty ITT

flow. The capacity limit for horizontal pipe at various pitches is given in Table 4-4.

The capacity of a steam pipe depends on these factors:

1. The quantity of condensate in the pipe.
2. The direction of condensate flow.
3. The pressure drop in the pipe. The total pressure drop of a system should not exceed the supply pressure when steam and condensate are flowing in the same direction.

When steam and condensate flow in the same direction, only the pressure drop need be considered. When steam flows counter to condensate, the velocities in Table 4-4 must not be exceeded. Table 4-5 gives steam-pipe capacities in square feet EDR when steam flow is counter to condensate flow in either one-pipe or two-pipe systems.

In all the pipe sizing tables given in this section, the *length of run* is defined as the *equivalent length of run*. The equivalent length consists of the measured length plus the allowance for fittings and valves which add resistance to the regular friction of the straight run of pipe. Table 4-6 gives the value in feet to be added to the measured length of pipe.

It is not possible at the beginning of the pipe sizing process to

**Table 4-2. Relations of Altitude, Pressure, and Boiling Point**

| Altitude Feet | ATMOSPHERIC PRESSURE ABSOLUTE | | BOILING POINT OF WATER °F (GAUGE PRESSURE, psi) | | | | |
|---|---|---|---|---|---|---|---|
| | Inches of Mercury (Barometer) | Lbs. per Sq. In. | 0 | 1 | 5 | 10 | 15 |
| −500 | 30.46 | 14.96 | 212.8 | 216.1 | 227.7 | 239.9 | 250.2 |
| −100 | 30.01 | 14.74 | 212.3 | 215.5 | 227.2 | 239.4 | 249.9 |
| Sea Level | 29.90 | 14.69 | 212.0 | 215.3 | 227.0 | 239.3 | 249.7 |
| 500 | 29.35 | 14.42 | 211.0 | 214.4 | 226.3 | 238.7 | 249.2 |
| 1000 | 28.82 | 14.16 | 210.1 | 213.5 | 225.5 | 238.1 | 248.6 |
| 1500 | 28.30 | 13.90 | 209.4 | 212.7 | 225.0 | 237.6 | 248.2 |
| 2000 | 27.78 | 13.65 | 208.2 | 211.7 | 224.1 | 236.8 | 247.7 |
| 2500 | 27.27 | 13.40 | 207.3 | 210.9 | 223.4 | 236.3 | 247.2 |
| 3000 | 26.77 | 13.15 | 206.4 | 210.1 | 222.7 | 235.7 | 246.7 |
| 3500 | 26.29 | 12.91 | 205.5 | 209.2 | 222.1 | 235.1 | 246.2 |
| 4000 | 25.81 | 12.68 | 204.7 | 208.4 | 221.4 | 234.6 | 245.7 |
| 4500 | 25.34 | 12.45 | 203.7 | 207.5 | 220.7 | 234.0 | 245.2 |
| 5000 | 24.88 | 12.22 | 202.7 | 206.8 | 220.1 | 233.4 | 244.7 |
| 6000 | 23.98 | 11.78 | 200.9 | 205.0 | 218.7 | 232.4 | 243.8 |
| 7000 | 23.11 | 11.35 | 199.1 | 203.3 | 217.3 | 231.3 | 242.9 |
| 8000 | 22.28 | 10.94 | 197.4 | 201.6 | 216.1 | 230.3 | 242.0 |
| 9000 | 21.47 | 10.55 | 195.7 | 200.0 | 214.8 | 229.3 | 241.3 |
| 10000 | 20.70 | 10.17 | 194.0 | 198.4 | 213.5 | 228.3 | 240.4 |
| 11000 | 19.95 | 9.80 | 192.2 | 196.8 | 212.3 | 227.3 | 239.6 |
| 12000 | 19.23 | 9.45 | 190.6 | 195.2 | 211.1 | 226.3 | 238.7 |
| 13000 | 18.53 | 9.10 | 188.7 | 193.6 | 209.9 | 225.4 | 237.9 |
| 14000 | 17.86 | 8.77 | 187.2 | 192.3 | 208.8 | 224.5 | 237.2 |
| 15000 | 17.22 | 8.46 | 185.4 | 190.6 | 207.6 | 223.6 | 236.4 |

Courtesy Hoffman Specialty ITT

## Table 4-3. Steam Capacity* of ASTM† Schedule 40(S) Pipe at Internal Pressures of 3.5 and 12 psig‡

Flow expressed in sq. ft. EDR

| NOMINAL PIPE SIZE inches | 1/16 psi (1 oz.) SAT. PR. 3.5 | 1/16 psi (1 oz.) psig 12 | 1/8 psi (2 oz.) SAT. PR. 3.5 | 1/8 psi (2 oz.) psig 12 | 1/4 psi (4 oz.) SAT. PR. 3.5 | 1/4 psi (4 oz.) psig 12 | 1/2 psi (8 oz.) SAT. PR. 3.5 | 1/2 psi (8 oz.) psig 12 | 3/4 psi (12 oz.) SAT. PR. 3.5 | 3/4 psi (12 oz.) psig 12 | 1 psi SAT. PR. 3.5 | 1 psi psig 12 | 2 psi SAT. PR. 3.5 | 2 psi psig 12 |
|---|---|---|---|---|---|---|---|---|---|---|---|---|---|---|
| 1 | 2 | 3 | 4 | 5 | 6 | 7 | 8 | 9 | 10 | 11 | 12 | 13 | 14 | 15 |
| 3/4" | 36 | 44 | 56 | 64 | 80 | 96 | 116 | 140 | 144 | 172 | 168 | 200 | 240 | 292 |
| 1" | 68 | 84 | 104 | 124 | 148 | 184 | 216 | 264 | 272 | 328 | 324 | 380 | 456 | 548 |
| 1 1/4" | 144 | 180 | 212 | 264 | 312 | 344 | 444 | 552 | 560 | 680 | 648 | 800 | 928 | 1120 |
| 1 1/2" | 224 | 280 | 326 | 400 | 480 | 588 | 696 | 840 | 872 | 1040 | 984 | 1216 | 1440 | 1720 |
| 2" | 432 | 536 | 648 | 776 | 936 | 1140 | 1344 | 1640 | 1680 | 2040 | 1920 | 2360 | 2840 | 3400 |
| 2 1/2" | 696 | 860 | 1032 | 1240 | 1512 | 1840 | 2160 | 2640 | 2720 | 3280 | 3120 | 3800 | 4600 | 5480 |
| 3" | 1272 | 1520 | 1860 | 2200 | 2640 | 3240 | 3840 | 4640 | 4760 | 5720 | 5520 | 6680 | 7800 | 9600 |
| 3 1/2" | 1848 | 2200 | 2680 | 3200 | 3960 | 4872 | 5640 | 6800 | 6960 | 8400 | 8000 | 9680 | 11800 | 13800 |
| 4" | 2560 | 3200 | 3800 | 4640 | 5640 | 6760 | 7920 | 9600 | 9800 | 12000 | 11520 | 13840 | 16800 | 19600 |
| 5" | 4800 | 5720 | 6720 | 8400 | 9760 | 12000 | 14280 | 17000 | 17520 | 21000 | 20400 | 24400 | 30000 | 34400 |
| 6" | 7680 | 9200 | 11280 | 13200 | 15840 | 19400 | 22800 | 28000 | 28800 | 34400 | 33600 | 40000 | 47600 | 56800 |
| 8" | 15600 | 19200 | 22280 | 28000 | 32400 | 40000 | 45600 | 57200 | 58000 | 70800 | 66000 | 82000 | 96000 | 118000 |
| 10" | 28800 | 35200 | 40800 | 50400 | 60000 | 72800 | 84000 | 104000 | 104800 | 128000 | 120000 | 148000 | 170800 | 208000 |
| 12" | 45600 | 54800 | 66000 | 78000 | 93600 | 113600 | 132000 | 160000 | 164000 | 198000 | 192000 | 230000 | 271200 | 324000 |

PRESSURE DROP—psi PER 100 FT. IN LENGTH

*Based on moody friction where flow of condensate does not inhibit the flow of steam.
†American Society for Testing Materials Schedule. The number 40 refers to the ASTM Schedule. The letter (S) refers to the former designation of standard weight pipe.
‡The flow rates at 3.5 psig can be used to cover saturated pressures from 1 to 6 psig, and the rates at 12 psig can be used to cover saturated pressures from 8 to 16 psig with an error not exceeding 8%.

Courtesy Hoffman Specialty ITT

## Table 4-4. Comparative Capacity of Steam Lines at Various Pitches for Steam and Condensate Flowing in Opposite Directions*

| Pitch of Pipe | PITCH OF PIPE IN INCHES PER 10 FT. / VELOCITY IN FT. PER SEC. / CAPACITY IN SQ. FT. EDR | | | | | | | | | | | | | | | |
|---|---|---|---|---|---|---|---|---|---|---|---|---|---|---|---|---|
| | ¼ Inch | | ½ Inch | | 1 Inch | | 1½ Inch | | 2 Inches | | 3 Inches | | 4 Inches | | 5 Inches | |
| PIPE SIZE inches | Capacity | Max. Vel. | Capacity | Max Vel. | Capacity | Max. Vel. | Capacity | Max. Vel | Capacity | Max. Vel. | Capacity | Max. Vel. | Capacity | Max. Vel. | Capacity | Max. Vel. |
| ¾ | 12.8 | 8 | 16.4 | 11 | 22.8 | 13 | 25.6 | 14 | 28.4 | 16 | 33.2 | 17 | 38.6 | 22 | 42.0 | 22 |
| 1 | 27.2 | 9 | 36.0 | 12 | 46.8 | 15 | 51.2 | 17 | 59.2 | 19 | 69.2 | 22 | 76.8 | 24 | 82.0 | 25 |
| 1¼ | 47.2 | 11 | 63.6 | 14 | 79.6 | 17 | 98.4 | 20 | 108.0 | 22 | 125.2 | 25 | 133.6 | 26 | 152.4 | 31 |
| 1½ | 79.2 | 12 | 103.6 | 16 | 132.0 | 19 | 149.6 | 22 | 168.0 | 24 | 187.2 | 26 | 203.2 | 28 | 236.8 | 33 |
| 2 | 171.6 | 15 | 216.0 | 18 | 275.2 | 24 | 333.2 | 27 | 371.6 | 30 | 398.4 | 32 | 409.6 | 32 | 460.0 | 33 |

* From research sponsored by ASHRAE.

Courtesy Hoffman Specialty ITT

### Table 4-5. Steam Pipe Capacities for Low Pressure Systems

(For use on one-pipe systems or two-pipe systems in
which condensate flows against the steam flow)

| | CAPACITY IN SQUARE FEET EDR | | | | |
|---|---|---|---|---|---|
| | TWO-PIPE SYSTEMS | | ONE-PIPE SYSTEMS | | |
| | Condensate Flowing against Steam | | Supply Risers Up-Feed | Radiator Valves and Vertical Connections | Radiator and Riser Horizontal Runouts |
| NOMINAL | VERTICAL | HORIZON-TAL | | | |
| A | B* | C† | D‡ | E | F† |
| ³/₄ | 32 | 28 | 24 | — | 28 |
| 1 | 56 | 56 | 44 | 28 | 28 |
| 1¼ | 124 | 108 | 80 | 64 | 64 |
| 1½ | 192 | 168 | 152 | 92 | 64 |
| 2 | 388 | 362 | 288 | 168 | 92 |
| 2½ | 636 | 528 | 464 | — | 168 |
| 3 | 1128 | 800 | 800 | — | 260 |
| 3½ | 1548 | 1152 | 1144 | — | 476 |
| 4 | 2044 | 1700 | 1520 | — | 744 |
| 5 | 4200 | 3152 | — | — | 1112 |
| 6 | 7200 | 5600 | — | — | 2180 |
| 8 | 15000 | 12000 | — | — | — |
| 10 | 28000 | 22800 | — | — | — |
| 12 | 46000 | 38000 | — | — | — |

*Do not use Col. B for pressure drops of less than 1 oz./100 ft. of equivalent length of run. Use Table 4-3 instead.

†Pitch of horizontal runouts to risers and radiators should not be less than ½ inch per ft. Where this pitch cannot be obtained, for runouts 8 ft. in length or over, increase one pipe size larger than shown in this table.

‡Do not use Col. D for pressure drops less than ²/₃ oz./100 ft. of equivalent length of run except for pipe size 3″ and over. Use Table 4-3 instead.

**Courtesy Hoffman Specialty ITT**

know the equivalent length of pipe so it is necessary to make an assumption. The accepted rule is to measure the actual length and double it to get the total equivalent length. After the pipe sizing is done for the proper pressure drop per 100′ of length based on the assumption, then a reexamination can be made to determine if any changes are required which will change the pipe sizing.

**Table 4-6. Frictional Allowance for Fittings in Feet of Pipe (to be Added to Actual Length of Run) Steam**

| SIZE OF PIPE inches | LENGTH IN FEET TO BE ADDED IN RUN | | | | |
|---|---|---|---|---|---|
| | STANDARD ELBOW | SIDE OUTLET TEE | GATE VALVE | GLOBE VALVE | ANGLE VALVE |
| ½" | 1.3 | 3 | 0.3 | 14 | 7 |
| ¾" | 1.8 | 4 | 0.4 | 18 | 10 |
| 1" | 2.2 | 5 | 0.5 | 23 | 12 |
| 1¼" | 3.0 | 6 | 0.6 | 29 | 15 |
| 1½" | 3.5 | 7 | 0.8 | 34 | 18 |
| 2" | 4.3 | 8 | 1.0 | 46 | 22 |
| 2½" | 5.0 | 11 | 1.1 | 54 | 27 |
| 3" | 6.5 | 13 | 1.4 | 66 | 34 |
| 3½" | 8 | 15 | 1.6 | 80 | 40 |
| 4" | 9 | 18 | 1.9 | 92 | 45 |
| 5" | 11 | 22 | 2.2 | 112 | 56 |
| 6" | 13 | 27 | 2.8 | 136 | 67 |
| 8" | 17 | 35 | 3.7 | 180 | 92 |
| 10" | 21 | 45 | 4.6 | 230 | 112 |

EXAMPLE.

| | | |
|---|---|---|
| MEASURED LENGTH | = | 132.0' |
| 4" GATE VALVE | = | 1.9' |
| 4'4" ELBOWS | = | 36.0' |
| EQUIV. LENGTH | | 169.9' |

—132'0"—

4"

Table 4-7. Return Main and Riser Capacities for Low Pressure Systems (sq. ft. EDR)

| PIPE SIZE INCHES | 1/32 psi or 1/2 oz. Drop per 100 ft. | | | 1/24 or 2/3 oz. Drop per 100 ft. | | | 1/16 psi or 1 oz. Drop per 100 ft. | | | 1/8 or 2 oz. Drop per 100 ft. | | | 1/4 psi or 4 oz. Drop per 100 ft. | | | 1/2 psi or 8 oz. Drop per 100 ft. | | |
|---|---|---|---|---|---|---|---|---|---|---|---|---|---|---|---|---|---|---|
| | WET | DRY | VAC | WET | DRY | VAC | WET | DRY | VAC | WET | DRY | VAC | WET | DRY | VAC | WET | DRY | VAC |
| G | H | I | J | K | L | M | N | O | P | Q | R | S | T | U | V | W | X | Y |
| **MAINS** | | | | | | | | | | | | | | | | | | |
| 3/4 | — | — | — | — | — | 168 | — | — | 400 | — | — | 568 | — | — | 800 | — | — | 1132 |
| 1 | 500 | 248 | — | 580 | 284 | 572 | 700 | 320 | 700 | 1000 | 412 | 996 | 1400 | 460 | 1400 | — | — | 1976 |
| 1 1/4 | 852 | 520 | — | 992 | 596 | 976 | 1200 | 672 | 1200 | 1700 | 868 | 1707 | 2400 | 964 | 2400 | — | — | 3392 |
| 1 1/2 | 1352 | 824 | — | 1572 | 944 | 1552 | 1900 | 1060 | 1900 | 2700 | 1360 | 2696 | 3800 | 1512 | 3800 | — | — | 5360 |
| 2 | 2800 | 1800 | — | 3240 | 2140 | 3260 | 4000 | 2300 | 4000 | 5600 | 2960 | 5680 | 8000 | 3300 | 8000 | — | — | 11320 |
| 2 1/2 | 4720 | 3040 | — | 6320 | 3472 | 5440 | 6720 | 3800 | 6720 | 9400 | 4920 | 9520 | 13400 | 5440 | 13400 | — | — | 18920 |
| 3 | 7520 | 5840 | — | 8520 | 6240 | 8720 | 10720 | 7000 | 10720 | 15000 | 9000 | 15200 | 21400 | 10000 | 21400 | — | — | 30240 |
| 3 1/2 | 11000 | 7880 | — | 13200 | 8800 | 13000 | 16000 | 10000 | 16000 | 22000 | 12920 | 22720 | 32000 | 14320 | 32000 | — | — | 45200 |
| 4 | 15520 | 11720 | — | 18320 | 13400 | 18000 | 22000 | 15000 | 22000 | 31000 | 19320 | 31240 | 44000 | 21520 | 44000 | — | — | 62000 |
| 5 | — | — | — | — | — | 31520 | — | — | 38720 | — | — | 54800 | — | — | 77600 | — | — | 109200 |
| 6 | — | — | — | — | — | 50400 | — | — | 62000 | — | — | 88000 | — | — | 124000 | — | — | 175200 |
| **RISERS** | | | | | | | | | | | | | | | | | | |
| 3/4 | — | — | — | — | — | 572 | — | — | 700 | — | — | 996 | — | — | 1400 | — | — | 1976 |
| 1 | — | 192 | — | — | 192 | 976 | — | 192 | 1200 | — | 192 | 1704 | — | 192 | 2400 | — | — | 3392 |
| 1 1/4 | — | 452 | — | — | 452 | 1552 | — | 452 | 1900 | — | 452 | 2696 | — | 452 | 3800 | — | — | 5360 |
| 1 1/2 | — | 992 | — | — | 992 | 3260 | — | 992 | 4000 | — | 992 | 5680 | — | 992 | 8000 | — | — | 11320 |
| 2 | — | 1500 | — | — | 1500 | 5400 | — | 1500 | 6720 | — | 1500 | 9520 | — | 1500 | 13400 | — | — | 18920 |
| 2 1/2 | — | 3000 | — | — | 3000 | 8720 | — | 3000 | 10720 | — | 3000 | 15200 | — | 3000 | 21400 | — | — | 30240 |
| 3 | — | — | — | — | — | 13000 | — | — | 16000 | — | — | 22720 | — | — | 32000 | — | — | 45200 |
| 3 1/2 | — | — | — | — | — | 17920 | — | — | 22000 | — | — | 31240 | — | — | 44000 | — | — | 62000 |
| 4 | — | — | — | — | — | 31520 | — | — | 38720 | — | — | 54800 | — | — | 77600 | — | — | 109200 |
| 5 | — | — | — | — | — | 50400 | — | — | 62000 | — | — | 88000 | — | — | 124000 | — | — | 175200 |

Courtesy Hoffman Specialty ITT

**Table 4-8. Horizontal Steam Supply Runouts to Radiator from Main or Riser and Radiator Supply Valve Sizes**

| | SIZE IN INCHES | | | | |
| | CAPACITY EDR | | | | |
| | 0-25 | 26-75 | 76-150 | 151-200 | 201-400 |
|---|---|---|---|---|---|
| Horizontal runout | 3/4 | 1 | 1 1/4 | 1 1/2 | 2 |
| Vertical connection | 1/2 | 3/4 | 1 | 1 1/4 | 1 1/2 |
| Supply valve | 1/2 | 3/4 | 1 | 1 1/4 | 1 1/2 |

**Table 4-9. Horizontal Return Runouts to Radiator from Main or Riser and Radiator Trap Size**

| | SIZE IN INCHES | | |
| | CAPACITY EDR | | |
| | 0-200 | 201-400 | 401-700 |
|---|---|---|---|
| Horizontal runout | 3/4 | 3/4 | 1 |
| Vertical connection | 1/2 | 3/4 | 1 |
| Trap | 1/2 | 3/4 | 1 |

Courtesy Hoffman Specialty ITT

## Table 4-10. Equivalent Capacities of Pipes of Same Length—Steam

| Size | NO. OF SMALL PIPES EQUIVALENT TO ONE LARGE PIPE | | | | | | | | | | | |
|---|---|---|---|---|---|---|---|---|---|---|---|---|
| | ½″ | ¾″ | 1″ | 1¼″ | 1½″ | 2″ | 2½″ | 3″ | 3½″ | 4″ | 5″ | 6″ |
| ½″ | 1.00 | 2.27 | 4.88 | 10.0 | 15.8 | 31.7 | 52.9 | 96.9 | 140 | 205 | 377 | 620 |
| ¾″ | | 1.00 | 2.05 | 4.30 | 6.97 | 14.0 | 23.3 | 42.5 | 65 | 90 | 166 | 273 |
| 1″ | | | 1.00 | 2.25 | 3.45 | 6.82 | 11.4 | 20.9 | 30 | 44 | 81 | 133 |
| 1¼″ | | | | 1.00 | 1.50 | 3.10 | 5.25 | 9.10 | 12 | 19 | 37 | 68 |
| 1½″ | | | | | 1.00 | 2.00 | 3.34 | 6.13 | 9 | 13 | 23 | 39 |
| 2″ | | | | | | 1.00 | 1.67 | 3.06 | 4.5 | 6.5 | 11.9 | 19.6 |
| 2½″ | | | | | | | 1.00 | 1.82 | 2.70 | 3.87 | 7.12 | 11.7 |
| 3″ | | | | | | | | 1.00 | 1.50 | 2.12 | 3.89 | 6.39 |
| 3½″ | | | | | | | | | 1.00 | 1.25 | 2.50 | 4.25 |
| 4″ | | | | | | | | | | 1.00 | 1.84 | 3.02 |
| 5″ | | | | | | | | | | | 1.00 | 1.65 |
| 6″ | | | | | | | | | | | | 1.00 |

This table may be used to find the number of smaller pipes equivalent in steam carrying capacity to one larger pipe. It may also be used to find the size of a larger pipe equivalent to several smaller ones. The pipes in either case must be of the same lengths.

*Example 1*—Find the number of 1″ pipes each 50′ long equivalent to one 4″ pipe 50′ long.

*Solution 1*—Follow down column headed 4″ to the point opposite 1″ in vertical column, and it is found that it will take 44 of the 1″ pipes in parallel to equal one 4″ pipe in steam carrying capacity.

*Example 2*—Find the size of one pipe equivalent to four 2″ pipes in steam carrying capacity.

*Solution 2*—Find 2″ in vertical column headed "Size" and follow across horizontally until closest number to 4 is found. The nearest to 4 is 4.5. Following this column up it is found that the size is 3½″. One pipe is, therefore, equivalent in steam carrying capacity to approximately four 2″ pipes.

## Table 4-11. Heat Losses from Covered Pipe

### (85 Percent Magnesia Type)

BTU PER LINEAR FOOT PER °F TEMPERATURE
DIFFERENCE (SURROUNDING AIR ASSUMED 75°F)

| Pipe Size | Insulation Thickness, Inches | MAX. TEMP. OF PIPE SURFACE °F. | | | | |
|---|---|---|---|---|---|---|
| | | 125 | 175 | 225 | 275 | 325 |
| ½ | 1 | .145 | .150 | .157 | .160 | .162 |
| ¾ | 1 | .165 | .172 | .177 | .180 | .182 |
| 1 | 1 | .190 | .195 | .200 | .203 | .207 |
| | 1½ | .160 | .165 | .167 | .170 | .175 |
| 1¼ | 1 | .220 | .225 | .232 | .237 | .245 |
| | 1½ | .182 | .187 | .193 | .197 | .200 |
| 1½ | 1 | .240 | .247 | .255 | .260 | .265 |
| | 1½ | .200 | .205 | .210 | .215 | .219 |
| 2 | 1 | .282 | .290 | .297 | .303 | .307 |
| | 1½ | .230 | .235 | .240 | .243 | .247 |
| | 2 | .197 | .200 | .205 | .210 | .217 |
| 2½ | 1 | .322 | .330 | .340 | .345 | .355 |
| | 1½ | .260 | .265 | .270 | .275 | .280 |
| | 2 | .220 | .225 | .230 | .237 | .242 |
| 3 | 1 | .375 | .385 | .395 | .405 | .415 |
| | 1½ | .300 | .305 | .312 | .320 | .325 |
| | 2 | .253 | .257 | .263 | .270 | .277 |
| 3½ | 1 | .419 | .430 | .440 | .450 | .460 |
| | 1½ | .332 | .339 | .345 | .352 | .360 |
| | 2 | 280 | .285 | .290 | .295 | .303 |
| 4 | 1 | .460 | .470 | .480 | .492 | .503 |
| | 1½ | .362 | .370 | .379 | .385 | .392 |
| | 2 | .303 | .308 | .315 | .320 | .327 |
| 5 | 1 | .545 | .560 | .572 | .585 | .600 |
| | 1½ | .423 | .435 | .442 | .450 | .460 |
| | 2 | .355 | .360 | .367 | .375 | .382 |
| 6 | 1 | .630 | .645 | .662 | .680 | .693 |
| | 1½ | .487 | .500 | .510 | .520 | .530 |
| | 2 | .405 | .415 | .420 | .430 | .437 |
| 8 | 1 | .790 | .812 | .835 | .850 | .870 |
| | 1½ | .603 | .620 | .635 | .645 | .660 |
| | 2 | .495 | .507 | .517 | .527 | .540 |

Courtesy Hoffman Specialty ITT

## Table 4-12. Heat Losses from Horizontal Bare Steel Pipe Sizes ³⁄₄ " to 6"

TEMP OF AIR SURROUNDING PIPE = 70°

BTU PERR HOUR PER LINEAR FOOT

PIPE TEMP °

Courtesy Hoffman Specialty ITT

# CHAPTER 5

# General and Useful Steam Heating Information

Tables 5-1 through 5-51, which follow, supply valuable reference data. Although these tables were prepared primarily for steam heating systems, they also contain information related to other piping and heating systems.

This chapter also contains definitions of common heating terms and a troubleshooting section.

## DEFINITIONS OF HEATING TERMS

The definitions given in this section are only those applying to heating and particularly as used in this book. Some do not define the terms for all usages, but in the interest of clearance and space this sacrifice was made.

## Table 5-1. Abbreviations Used in Heating

| | |
|---|---|
| Absolute . . . . . . . . . . . . . . . abs | Gallons per Minute . . . . . . . gpm |
| Alternating-Current . . . . . . . . a-c | Gallons per Second . . . . . . . . gps |
| Ampere . . . . . . . . . . . . . . . amp | Gram . . . . . . . . . . . . . . . . . . . . g |
| Atmosphere . . . . . . . . . . . . . atm | Horsepower . . . . . . . . . . . . . . hp |
| Average . . . . . . . . . . . . . . . avg | Horsepower-Hour . . . . . . . . hp hr |
| Avoirdupois . . . . . . . . . . . . . avdp | Hour . . . . . . . . . . . . . . . . . . . . hr |
| Barometer . . . . . . . . . . . . . . bar. | Inch . . . . . . . . . . . . . . . . . . . . in. |
| Boiling Point . . . . . . . . . . . . . bp | Inch-Pound . . . . . . . . . . . . . in.-lb |
| Brake Horsepower . . . . . . . . . bhp | Kilogram . . . . . . . . . . . . . . . . kg |
| Brake Horsepower-Hour . . . bhp-hr | Kilowatt . . . . . . . . . . . . . . . . kw |
| British Thermal Unit . . . . . . . Btu | Melting Point. . . . . . . . . . . . . mp |
| British Thermal Units | Meter . . . . . . . . . . . . . . . . . . . m |
|    per Hour . . . . . . . . . . . . . Btuh | Miles per Hour . . . . . . . . . . mph |
| Calorie . . . . . . . . . . . . . . . . cal | Millimeter . . . . . . . . . . . . . . mm |
| Centigram . . . . . . . . . . . . . . . cg | Minute . . . . . . . . . . . . . . . . . mm |
| Centimeter . . . . . . . . . . . . . . cm | Ounce. . . . . . . . . . . . . . . . . . . oz |
| Cubic . . . . . . . . . . . . . . . . . . cu | Pound . . . . . . . . . . . . . . . . . . lb |
| Cubic Centimeter . . . . . . . . . . cc | Pounds per Square Inch. . . . . psi |
| Cubic Foot . . . . . . . . . . . . . . cu ft | Pounds per Square Inch, |
| Cubic Feet per Minute . . . . . . cfm |    Gauge . . . . . . . . . . . . . . psig |
| Cubic Feet per Second . . . . . . cfs | Pounds per Square Inch, |
| Degree . . . . . . . . . . . . . deg or ° |    Absolute . . . . . . . . . . . . . psia |
| Degree, Centigrade . . . . . . . . C | Revolutions per Minute . . . . . rpm |
| Degree, Fahrenheit . . . . . . . . . F | Revolutions per Second . . . . . rps |
| Diameter . . . . . . . . . . . . . . . diam | Second . . . . . . . . . . . . . . . . sec |
| Direct-Current . . . . . . . . . . . . d-c | Specific Gravity . . . . . . . . . . sp gr |
| Feet per Minute. . . . . . . . . . . fpm | Specific Heat . . . . . . . . . . . . sp ht |
| Feet per Second . . . . . . . . . . fps | Square Foot . . . . . . . . . . . . . sq ft |
| Foot . . . . . . . . . . . . . . . . . . . ft | Square Inch . . . . . . . . . . . sq in. |
| Foot-Pound . . . . . . . . . . . . . ft-lb | Volt. . . . . . . . . . . . . . . . . . . . . v |
| Freezing Point . . . . . . . . . . . . fp | Watt . . . . . . . . . . . . . . . . . . . w |
| Gallon. . . . . . . . . . . . . . . . . gal | Watt Hour . . . . . . . . . . . . . . whr |

*Absolute Humidity*—The weight of water vapor in grains actually contained in one cubic foot of the mixture of air and moisture.

*Absolute Pressure*—The actual pressure above zero. It is the atmospheric pressure added to the gauge pressure. It is also expressed as a unit pressure, such as lbs. per sq. in.

*Absolute Temperature*—The temperature of a substance measured above absolute zero. To express a temperature as absolute temperature add 460° to the reading of a Fahrenheit thermometer or 273° to the reading of a Centigrade one.

*Absolute Zero*—The temperature ( − 460°F approx.) at which all molecular motion of a substance ceases, and at which the substance contains no heat.

*Air*—An elastic gas. It is a mechanical mixture of oxygen and nitrogen and slight traces of other gases. It may also contain moisture known as humidity. Dry air weighs 0.075 lbs. per cu. ft. One Btu will raise the temperature of 55 cu. ft. of air 1°F.

**Table 5-2. Inside Temperatures Usually Specified**

| BUILDING | ROOM | DEG. F. |
|---|---|---|
| SCHOOLS | Class Rooms | 70-72 |
| | Assembly Rooms | 61-72 |
| | Gymnasiums | 55-65 |
| | Toilets and Baths | 70 |
| | Wardrobe and Locker Rooms | 65-68 |
| | Kitchens | 66 |
| | Dining and Lunch Rooms | 65-70 |
| | Play Rooms | 60-65 |
| | Natatoriums | 75 |
| THEATRES | Seating Space | 68-72 |
| | Lounge Rooms | 68-70 |
| | Toilets | 68 |
| HOSPITALS | Private Rooms | 70-72 |
| | Private Rooms (Surgical) | 70-80 |
| | Operating Rooms | 70-95 |
| | Wards | 68 |
| | Kitchens and Laundries | 66 |
| | Toilets | 68 |
| | Bath Rooms | 70-80 |
| HOTELS | Bed Rooms and Baths | 70 |
| | Dining Rooms | 70 |
| | Kitchens and Laundries | 66 |
| | Ball Rooms | 65-68 |
| | Toilets and Service Rooms | 68 |
| HOMES | | 70-72 |
| Stores | | 65-68 |
| Public Buildings | | 68-72 |
| Warm Air Baths | | 120 |
| Steam Baths | | 110 |
| Factories and Machine Shops | | 60-65 |
| Foundries and Boiler Shops | | 50-60 |
| Paint Shops | | 80 |

## Table 5-3. Heating and Ventilating Symbols

HIGH PRESSURE STEAM SUPPLY PIPE

LOW PRESSURE STEAM SUPPLY PIPE

HOW WATER PIPE-FLOW (OR WET RETURN)

RETURN PIPE-STEAM OR WATER (OR DRY RETURN)

AIR VENT LINE

| | |
|---|---|
| FLANGES | COLUMN RADIATOR (PLAN) |
| SCREWED UNION | COLUMN RADIATOR (ELEVATION) |
| ELBOW | COLUMN RADIATOR (END VIEW) |
| ELBOW LOOKING UP | |
| ELBOW LOOKING DOWN | WALL RADIATOR (PLAN) |
| TEE | WALL RADIATOR (ELEVATION) |
| TEE LOOKING UP | |
| TEE LOOKING DOWN | WALL RADIATOR (END VIEW) |
| GATE VALVE | |
| GLOBE VALVE | PIPE COIL (PLAN) |
| ANGLE VALVE | PIPE COIL (ELEVATION) |
| ANGLE VALVE (STEM PERPENDICULAR) | |
| LOCK SHEILD VALVE | PIPE COIL (END VIEW) |
| CHECK VALVE | INDIRECT RADIATOR (PLAN) |
| REDUCING VALVE | INDIRECT RADIATOR (ELEVATION) |
| DIAPHRAGM VALVE | INDIRECT RADIATOR (END VIEW) |
| DIAPHRAGM VALVE (STEM PERPENDICULAR) | SUPPLY DUCT (SECTION) |
| THERMOSTAT | EXHAUST DUCT (SECTION) |
| RADIATOR TRAP (ELEVATION) | BUTTERFLY DAMPER (PLAN OR ELEVATION) |
| RADIATOR TRAP (PLAN) | |
| EXPANSION JOINT | BUTTERFLY DAMPER (ELEVATION OR PLAN) |
| AIR SUPPLY OUTLET | DEFLECTING DAMPER (SQUARE PIPE) |
| EXHAUST OUTLET | VANES |

**Courtesy Hoffman Specialty ITT**

## Table 5-4. Air Changes to Be Provided for Under Average Conditions Exclusive of Air Required for Ventilation

| KIND OF ROOM OR BUILDING | Number of Changes Taking Place Per Hour |
|---|---|
| Rooms, 1—Side Exposed | 1 |
| Rooms, 2—Sides Exposed | $1\frac{1}{2}$ |
| Rooms, 3—Sides Exposed | 2 |
| Rooms, 4—Sides Exposed | 2 |
| Rooms with No Windows or Outside Doors | $\frac{1}{2}$ to $\frac{3}{4}$ |
| Entrance Halls | 2 to 3 |
| Reception Halls | 2 |
| Living Rooms | 1 to 2 |
| Dining Rooms | 1 to 2 |
| Bath Rooms | 2 |
| Drug Stores | 2 to 3 |
| Clothing Stores | 1 |
| Churches, Factories, Lofts, etc. | $\frac{1}{2}$ to 3 |

Add 1 air change to above for rooms with a fireplace. While the most accurate way of estimating air infiltration considers leakage through walls and clearance between window frames and sash, the table above is sufficiently accurate for practical purposes. Where windows are equipped with metal weatherstripping, use ½ air change less per hour than shown above. For example: A room with 2 sides exposed, windows weatherstripped, would be figured at 1 air change instead of 1½.

### EXTRA SOURCES OF HEAT

In auditoriums of churches or theaters, or in factories where a large number of people are gathered, a considerable amount of heat is liberated and should be subtracted from the Btu required to heat the room. A person at rest gives off about 400 Btu/hour, while one at work gives off somewhere between 500 and 800 Btu/hour.

Electricity used in the room may be quite an item. The total wattage of electricity actually used multiplied by 3.4 should be subtracted from the required heating load (Btu/hour). In factories where gas or oil is used as in process work, the Btu/hour consumed should be subtracted from the required heating load. The heating valves of various kinds of gas and oil may be found elsewhere in the book.

**Courtesy Hoffman Specialty ITT**

Air expands or contracts approximately ¼₉₀ of its volume for each degree rise or fall in temperature from 32°F.

*Air Change*—The number of times in an hour the air in a room is changed either by mechanical means or by the infiltration of outside air leaking into the room through cracks around doors and windows.

*Air Cleaner*—A device designed for the purpose of removing airborne impurities, such as dust, fumes, and smokes. (Air cleaners include air washers and air filters.)

*Air Conditioning*—The simultaneous control of the temperature, humidity, air motion, and air distribution within an enclosure. Where comfort and health are involved, a reasonable air

## Table 5-5. Stock Assemblies of Small-Tube Cast-Iron Radiators

| Number of Sections | RATING, IN SQUARE FEET | | | | | | | |
|---|---|---|---|---|---|---|---|---|
| | 3 TUBE | 4 TUBE | | 5 TUBE | | 6 TUBE | | |
| | 25 in. | 25 in. | 22 in. | 25 in. | 22 in. | 32 in. | 25 in. | 19 in. |
| 6 | 9.6 | 12.0 | 10.8 | 14.4 | 12.6 | 22.2 | 18.0 | — |
| 10 | 16.0 | 20.0 | 18.0 | 24.0 | 21.0 | 37.0 | 30.0 | 23.0 |
| 14 | 22.4 | 28.0 | 25.2 | 33.6 | 29.4 | 51.8 | 42.0 | 32.2 |
| 18 | 28.8 | 36.0 | 32.4 | 43.2 | 37.8 | 66.6 | 54.0 | 41.4 |
| 22 | 35.2 | 44.0 | 39.6 | 52.8 | 46.2 | 81.4 | 66.0 | 50.6 |
| 26 | 41.6 | 52.0 | 46.8 | 62.4 | 54.6 | 96.2 | 78.0 | 59.8 |
| 30 | 48.0 | 60.0 | 54.0 | 7.20 | 63.0 | — | 90.0 | 69.0 |
| 38 | 60.8 | 76.0 | 68.4 | 91.2 | 79.8 | — | — | 87.4 |

**NIPPLES**—Sections are assembled with cast- or malleable-iron push nipples.

**TAPPINGS**—Tappings are standard iron-pipe size. Flow and return tappings are located horizontally opposite the top and bottom nipple ports. Air-vent tappings for water and steam radiators are provided on the end section opposite the supply section.

**PIPE FITTINGS**—Iron pipe size plugs or bushings, or both, may be furnished with each radiator.

**PAINTING**—Each section or radiator assembled by the manufacturer is given one priming coat.

**NUMBER OF SECTIONS**—The stock assemblies are shown in the table. When assemblies of more sections than those listed are required, the maximum number should not exceed 56, to avoid damage in shipping and handling. Consult manufacturers' catalogs.

The square foot equivalent direct steam radiation is defined as the ability to emit 240 Btu per hour, with steam at 215°F in air at 70°F. These ratings apply only to radiators installed exposed in a normal manner, not to radiators installed behind enclosures, grilles, and so on.

### Table 5-5. Stack Assemblies
### of Small-Tube Cast-Iron Radiators (Continued)

| Number of Tubes per Section | Catalog Rating per Section Sq. Ft. | SECTION DIMENSIONS IN INCHES | | | | |
|---|---|---|---|---|---|---|
| | | A HEIGHT* | B WIDTH | | C SPACING | D LEG HEIGHT* |
| | | | Min. | Max. | | |
| 3 | 1.6 | 26 | $3^1/_4$ | $3^1/_2$ | $1^3/_4$ | $2^1/_2$ |
| 4 | 1.6 | 19 | $4^7/_{16}$ | $4^{13}/_{16}$ | $1^3/_4$ | $2^1/_2$ |
| 4 | 1.8 | 22 | $4^7/_{16}$ | $4^{13}/_{16}$ | $1^3/_4$ | $2^1/_2$ |
| 4 | 2.0 | 25 | $4^7/_{16}$ | $4^{13}/_{16}$ | $1^3/_4$ | $2^1/_2$ |
| 5 | 2.1 | 22 | $5^5/_8$ | $6^5/_{16}$ | $1^3/_4$ | $2^1/_2$ |
| 5 | 2.4 | 25 | $5^5/_8$ | $6^5/_{16}$ | $1^3/_4$ | $2^1/_2$ |
| 6 | 2.3 | 19 | $6^{13}/_{16}$ | 8 | $1^3/_4$ | $2^1/_2$ |
| 6 | 3.0 | 25 | $6^{13}/_{16}$ | 8 | $1^3/_4$ | $2^1/_2$ |
| 6 | 3.7 | 32 | $6^{13}/_{16}$ | 8 | $1^3/_4$ | $2^1/_2$ |

*Overall height and leg height of radiator as made by some manufacturers is 1"greater than shown in Columns A and D. Radiators may be furnished without legs. Where greater than standard leg heights are required, this dimension is to be $4^1/_2$".

Courtesy Hoffman Specialty ITT

purity with regard to dust, bacteria, and odors is also included. The primary requirement of a good air-conditioning system is a good heating system.

*Air Infiltration*—The leakage of air into a house through cracks and crevices, and through doors, windows, and other openings, caused by wind pressure and/or temperature difference.

*Air Valve*—See Vent Valve.

*Atmospheric Pressure*—The weight of a column of air, 1 square inch in cross section and extending from the earth to the upper level of the blanket of air surrounding the earth. This air exerts a pressure of 14.7 lbs. per square inch at sea level, where water will boil at 212°F. High altitudes have lower atmospheric pressure with correspondingly lower boiling-point temperatures.

*Boiler*—A closed vessel in which steam is generated or in which water is heated by fire.

*Boiler Heating Surface*—The area of the heat transmitting surfaces in contact with the water (or steam) in the boiler on one side and the fire or hot gases on the other.

*Boiler Horsepower*—The equivalent of 34.5 lbs. of water per hour at 212°F to steam at 212°F. This is equal to a heat output of

167

## Table 5-6. Output of Direct Cast-Iron Radiators[a]

| STEAM PRESSURE (APPROXIMATE) | | STEAM OR MEAN WATER TEMP. °F | BTU PER SQ. FT. EDR. PER HOUR ROOM TEMPERATURE °F | | | | | | |
|---|---|---|---|---|---|---|---|---|---|
| GAUGE[b] | ABSOLUTE[c] Lbs. per Sq. In. | | 80 | 75 | 70 | 65 | 60 | 55 | 50 |
| **VACUUM Inches of MERCURY** | | | | | | | | | |
| 22.4 | 3.7 | 150 | 93 | 102 | 111 | 120 | 129 | 139 | 148 |
| 20.3 | 4.7 | 160 | 111 | 120 | 129 | 139 | 148 | 158 | 167 |
| 17.7 | 6.0 | 170 | 129 | 139 | 148 | 158 | 167 | 178 | 188 |
| 14.6 | 7.5 | 180 | 148 | 158 | 167 | 178 | 188 | 198 | 209 |
| 10.9 | 9.3 | 190 | 167 | 178 | 188 | 198 | 209 | 218 | 229 |
| 6.5 | 11.5 | 200 | 188 | 198 | 209 | 218 | 229 | 240 | 250 |
| 3.9 | 12.8 | 205 | 198 | 209 | 218 | 229 | 240 | 250 | 261 |
| 0.0 | 14.7 | 212 | 211 | 222 | 233 | 242 | 253 | 264 | 276 |
| 1. | 15.6 | 215 | 218 | 229 | 240 | 250 | 261 | 273 | 282 |
| 2. | 17 | 220 | 229 | 240 | 250 | 261 | 273 | 282 | 296 |
| **Lbs. per Sq. Inch** | | | | | | | | | |
| 6. | 21. | 230 | 250 | 261 | 273 | 282 | 296 | 308 | 316 |
| 10. | 25. | 240 | 273 | 282 | 296 | 308 | 316 | 329 | 343 |
| 15. | 30. | 250 | 296 | 308 | 316 | 329 | 343 | 353 | 364 |
| 27. | 42. | 270 | 343 | 353 | 364 | 375 | 387 | 400 | 414 |
| 52. | 67. | 300 | 414 | 421 | 436 | 453 | 462 | 471 | 490 |

[a]These outputs also apply quite closely to the output of the "R" type cast iron radiant baseboards. For exact outputs, the catalogs of the manufacturers of the baseboards should be consulted.

[b]At sea level only.

[c]At locations other than sea level use temperature only or convert gauge reading to absolute pressure. Add gauge reading to atmospheric pressure in lbs. per sq. in. for given altitude. To convert vacuum (inches of mercury) to absolute, multiply inches vacuum by 0.49 and deduct from atmospheric pressure (lbs. per sq. in.) for given altitude.

Courtesy Hoffman Specialty ITT

## Table 5-7. Standard Column Radiation*

### SINGLE-COLUMN RADIATORS — HEATING SURFACE-SQUARE FEET

| No. of Sections | Length 2½"-Sec. | 38" 3 Sq. Ft. Sec. | 32" 2½ Sq. Ft. Sec. | 26" 2 Sq. Ft. Sec. | 23" 1⅔ Sq. Ft. Sec. | 20" 1½ Sq. Ft. Sec. |
|---|---|---|---|---|---|---|
| 3 | 7½ | 9 | 7½ | 6 | 5 | 4½ |
| 4 | 10 | 12 | 10 | 8 | 6⅔ | 6 |
| 5 | 12½ | 15 | 12½ | 10 | 8⅓ | 7½ |
| 6 | 15 | 18 | 15 | 12 | 10 | 9 |
| 7 | 17½ | 21 | 17½ | 14 | 11⅔ | 10½ |
| 8 | 20 | 24 | 20 | 16 | 13⅓ | 12 |
| 9 | 22½ | 27 | 22½ | 18 | 15 | 13½ |
| 10 | 25 | 30 | 25 | 20 | 16⅔ | 15 |
| 11 | 27½ | 33 | 27½ | 22 | 18⅓ | 16½ |
| 12 | 30 | 36 | 30 | 24 | 20 | 18 |
| 13 | 32½ | 39 | 32½ | 26 | 21⅔ | 19½ |
| 14 | 35 | 42 | 35 | 28 | 23⅓ | 21 |
| 15 | 37½ | 45 | 37½ | 30 | 25 | 22½ |
| 16 | 40 | 48 | 40 | 32 | 26⅔ | 24 |
| 17 | 42½ | 51 | 42½ | 34 | 28⅓ | 25½ |
| 18 | 45 | 54 | 45 | 36 | 30 | 27 |
| 19 | 47½ | 57 | 47½ | 38 | 31⅓ | 28½ |
| 20 | 50 | 60 | 50 | 40 | 33⅓ | 30 |
| 21 | 52½ | 63 | 52½ | 42 | 35 | 31½ |
| 22 | 55 | 66 | 55 | 44 | 36⅔ | 33 |
| 23 | 57½ | 69 | 57½ | 46 | 38⅓ | 34½ |
| 24 | 60 | 72 | 60 | 48 | 40 | 36 |
| 25 | 62½ | 75 | 62½ | 50 | 41⅓ | 37½ |
| 26 | 65 | 78 | 65 | 52 | 43⅓ | 39 |
| 27 | 67½ | 81 | 67½ | 54 | 45 | 40½ |

### TWO-COLUMN RADIATORS — HEATING SURFACE-SQUARE FEET

| No. of Sections | Length 2½"-Sec. | 45" 5 Sq. Ft. Sec. | 38" 4 Sq. Ft. Sec. | 32" 3⅓ Sq. Ft. Sec. | 26" 2⅔ Sq. Ft. Sec. | 23" 2⅓ Sq. Ft. Sec. | 20" 2 Sq. Ft. Sec. |
|---|---|---|---|---|---|---|---|
| 3 | 7½ | 15 | 12 | 10 | 8 | 7 | 6 |
| 4 | 10 | 20 | 16 | 13⅓ | 10⅔ | 9⅓ | 8 |
| 5 | 12½ | 25 | 20 | 16⅔ | 13⅓ | 11⅔ | 10 |
| 6 | 15 | 30 | 24 | 20 | 16 | 14 | 12 |
| 7 | 17½ | 35 | 28 | 23⅓ | 18⅔ | 16⅓ | 14 |
| 8 | 20 | 40 | 32 | 26⅔ | 21⅓ | 18⅔ | 16 |
| 9 | 22½ | 45 | 36 | 30 | 24 | 21 | 18 |
| 10 | 25 | 50 | 40 | 33⅓ | 26⅔ | 23⅓ | 20 |
| 11 | 27½ | 55 | 44 | 36⅔ | 29⅓ | 25⅔ | 22 |
| 12 | 30 | 60 | 48 | 40 | 33 | 28 | 24 |
| 13 | 32½ | 65 | 52 | 43⅓ | 34⅔ | 30⅓ | 26 |
| 14 | 35 | 70 | 56 | 46⅔ | 37⅓ | 32⅔ | 28 |
| 15 | 37½ | 75 | 60 | 50 | 40 | 35 | 30 |
| 16 | 40 | 80 | 64 | 53⅓ | 42⅔ | 37⅓ | 32 |
| 17 | 42½ | 85 | 68 | 56⅔ | 45⅓ | 39⅔ | 34 |
| 18 | 45 | 90 | 72 | 60 | 48 | 42 | 36 |
| 19 | 47½ | 95 | 76 | 63⅓ | 50⅔ | 44⅓ | 38 |
| 20 | 50 | 100 | 80 | 66⅔ | 53⅓ | 46⅔ | 40 |
| 21 | 52½ | 105 | 84 | 70 | 56 | 49 | 42 |
| 22 | 55 | 110 | 88 | 73⅓ | 58⅔ | 51⅓ | 44 |
| 23 | 57½ | 115 | 92 | 76⅔ | 61⅓ | 53⅔ | 46 |
| 24 | 60 | 120 | 96 | 80 | 64 | 56 | 48 |
| 25 | 62½ | 125 | 100 | 83⅓ | 66⅔ | 58⅓ | 50 |

Courtesy Hoffman Specialty ITT

*This table covers column radiation manufactured prior to 1926.

169

## Table 5-8. Standard Column Radiation*

### THREE-COLUMN RADIATORS — HEATING SURFACE—SQUARE FEET

| No. of Sections | Length 2½"-Sec. | 45" 6 Sq. Ft. Sec. | 38" 5 Sq. Ft. Sec. | 32" 4½ Sq. Ft. Sec. | 26" 3¼ Sq. Ft. Sec. | 22" 3 Sq. Ft. Sec. | 18" 2¼ Sq. Ft. Sec. |
|---|---|---|---|---|---|---|---|
| 3 | 7½ | 18 | 15 | 13½ | 11¼ | 9 | 6¾ |
| 4 | 10 | 24 | 20 | 18 | 15 | 12 | 9 |
| 5 | 12½ | 30 | 25 | 22½ | 18¾ | 15 | 11¼ |
| 6 | 15 | 36 | 30 | 27 | 22½ | 18 | 13½ |
| 7 | 17½ | 42 | 35 | 31½ | 26¼ | 21 | 15¾ |
| 8 | 20 | 48 | 40 | 36 | 30 | 24 | 18 |
| 9 | 22½ | 54 | 45 | 40½ | 33¾ | 27 | 20¼ |
| 10 | 25 | 60 | 50 | 45 | 37½ | 30 | 22½ |
| 11 | 27½ | 66 | 55 | 49½ | 41¼ | 33 | 24¾ |
| 12 | 30 | 72 | 60 | 54 | 45 | 36 | 27 |
| 13 | 32½ | 78 | 65 | 58½ | 48¾ | 39 | 29¼ |
| 14 | 35 | 84 | 70 | 63 | 52½ | 42 | 31½ |
| 15 | 37½ | 90 | 75 | 67½ | 56¼ | 45 | 33¾ |
| 16 | 40 | 96 | 80 | 72 | 60 | 48 | 36 |
| 17 | 42½ | 102 | 85 | 76½ | 63¾ | 51 | 38¼ |
| 18 | 45 | 108 | 90 | 81 | 67½ | 54 | 40½ |
| 19 | 47½ | 114 | 95 | 85½ | 71¼ | 57 | 42¾ |
| 20 | 50 | 120 | 100 | 90 | 75 | 60 | 45 |
| 21 | 52½ | 126 | 105 | 94½ | 78¾ | 63 | 47¼ |
| 22 | 55 | 132 | 110 | 99 | 82½ | 66 | 49½ |
| 23 | 57½ | 138 | 115 | 103½ | 86¼ | 69 | 51¾ |
| 24 | 60 | 144 | 120 | 108 | 90 | 72 | 54 |
| 25 | 62½ | 150 | 125 | 112½ | 93¾ | 75 | 56¼ |
| 26 | 65 | 156 | 130 | 117 | 97½ | 78 | 58½ |
| 27 | 67½ | 162 | 135 | 121½ | 101¼ | 81 | 60¾ |

### FOUR-COLUMN RADIATORS — HEATING SURFACE—SQUARE FEET

| No. of Sections | Length 3"-Sec. | 45" 10 Sq. Ft. Sec. | 38" 8 Sq. Ft. Sec. | 32" 6½ Sq. Ft. Sec. | 26" 6 Sq. Ft. Sec. | 22" 4 Sq. Ft. Sec. | 18" 3 Sq. Ft. Sec. |
|---|---|---|---|---|---|---|---|
| 3 | 9 | 30 | 24 | 19½ | 15 | 12 | 9 |
| 4 | 12 | 40 | 32 | 26 | 20 | 16 | 12 |
| 5 | 15 | 50 | 40 | 32½ | 25 | 20 | 15 |
| 6 | 18 | 60 | 48 | 39 | 30 | 24 | 18 |
| 7 | 21 | 70 | 56 | 45½ | 35 | 28 | 21 |
| 8 | 24 | 80 | 64 | 52 | 40 | 32 | 24 |
| 9 | 27 | 90 | 72 | 58½ | 45 | 36 | 27 |
| 10 | 30 | 100 | 80 | 65 | 50 | 40 | 30 |
| 11 | 33 | 110 | 88 | 71½ | 55 | 44 | 33 |
| 12 | 36 | 120 | 96 | 78 | 60 | 48 | 36 |
| 13 | 39 | 130 | 104 | 84½ | 65 | 52 | 39 |
| 14 | 42 | 140 | 112 | 91 | 70 | 56 | 42 |
| 15 | 45 | 150 | 120 | 97½ | 75 | 60 | 45 |
| 16 | 48 | 160 | 128 | 104 | 80 | 64 | 48 |
| 17 | 51 | 170 | 136 | 110½ | 85 | 68 | 51 |
| 18 | 54 | 180 | 144 | 117 | 90 | 72 | 54 |
| 19 | 57 | 190 | 152 | 123½ | 95 | 76 | 57 |
| 20 | 60 | 200 | 160 | 130 | 100 | 80 | 60 |
| 21 | 63 | 210 | 168 | 136½ | 105 | 84 | 63 |
| 22 | 66 | 220 | 176 | 143 | 110 | 88 | 66 |
| 23 | 69 | 230 | 184 | 149½ | 115 | 92 | 69 |
| 24 | 72 | 240 | 192 | 156 | 120 | 96 | 72 |
| 25 | 75 | 250 | 200 | 162½ | 125 | 100 | 75 |

Courtesy Hoffman Specialty ITT

*This table covers column radiation prior to 1926.

33,475 Btu/hr, which is equal to approximately 140 sq ft. of steam radiation (EDR).

*British Thermal Unit (Btu)*—The quantity of heat required to raise the temperature of 1 lb. of water 1°F. This is somewhat approximate but sufficiently accurate for any work discussed in this book.

*Bucket Trap (Inverted)*—A float trap with an open float. The float or bucket is open at the bottom. When the air or steam in the

### Table 5-9. Heating Surface—Three-Tube Radiators for Steam or Hot Water

| No. of Sections | Length Inches | HEATING SURFACE—SQUARE FEET 240 Btu PER Sq. Ft. PER HOUR | | | | |
|---|---|---|---|---|---|---|
| | | 36" HEIGHT 3½ Sq. Ft. Per Sec. | 30" HEIGHT 3 Sq. Ft. Per Sec. | 26" HEIGHT 2⅓ Sq. Ft. Per Sec. | 23" HEIGHT 2 Sq. Ft. Per Sec. | 20" HEIGHT 1¾ Sq. Ft. Per Sec. |
| 2 | 5 | 7 | 6 | 4⅔ | 4 | 3½ |
| 3 | 7½ | 10½ | 9 | 7 | 6 | 5¼ |
| 4 | 10 | 14 | 12 | 9⅓ | 8 | 7 |
| 5 | 12½ | 17½ | 15 | 11⅔ | 10 | 8¾ |
| 6 | 15 | 21 | 18 | 14 | 12 | 10½ |
| 7 | 17½ | 24½ | 21 | 16⅓ | 14 | 12¼ |
| 8 | 20 | 28 | 24 | 18⅔ | 16 | 14 |
| 9 | 22½ | 31½ | 27 | 21 | 18 | 15¾ |
| 10 | 25 | 35 | 30 | 23⅓ | 20 | 17½ |
| 11 | 27½ | 38½ | 33 | 25⅔ | 22 | 19¼ |
| 12 | 30 | 42 | 36 | 28 | 24 | 21 |
| 13 | 32½ | 45½ | 39 | 30⅓ | 26 | 22¾ |
| 14 | 35 | 49 | 42 | 32⅔ | 28 | 24½ |
| 15 | 37½ | 52½ | 45 | 35 | 30 | 26¼ |
| 16 | 40 | 56 | 48 | 37⅓ | 32 | 28 |
| 17 | 42½ | 59½ | 51 | 39⅔ | 34 | 29¾ |
| 18 | 45 | 63 | 54 | 42 | 36 | 31½ |
| 19 | 47½ | 66½ | 57 | 44⅓ | 38 | 33¼ |
| 20 | 50 | 70 | 60 | 46⅔ | 40 | 35 |
| 21 | 52½ | 73½ | 63 | 49 | 42 | 36¾ |
| 22 | 55 | 77 | 66 | 51⅓ | 44 | 38½ |
| 23 | 57½ | 80½ | 69 | 53⅔ | 46 | 40¼ |
| 24 | 60 | 84 | 72 | 56 | 48 | 42 |
| 25 | 62½ | 87½ | 75 | 58⅓ | 50 | 43¾ |

NOTE—These data apply to obsolete radiators manufactured since 1926 but not manufactured today.

Courtesy Hoffman Specialty ITT

## Table 5-10. Heating Surface—Four-Tube Radiators
### for Steam or Hot Water

| No. of Sections | Length Inches | HEATING SURFACE—SQUARE FEET 240 Btu PER SQ. FT. PER HOUR | | | | |
|---|---|---|---|---|---|---|
| | | 37" HEIGHT 4¼ Sq. Ft. Per Sec. | 32" HEIGHT 3½ Sq. Ft. Per Sec. | 26" HEIGHT 2¾ Sq. Ft. Per Sec. | 23" HEIGHT 2½ Sq. Ft. Per Sec. | 20" HEIGHT 2¼ Sq. Ft. Per Sec. |
| 2 | 5 | 8½ | 7 | 5½ | 5 | 4½ |
| 3 | 7½ | 12¾ | 10½ | 8¼ | 7½ | 6¾ |
| 4 | 10 | 17 | 14 | 11 | 10 | 9 |
| 5 | 12½ | 21¼ | 17½ | 13¾ | 12½ | 11¼ |
| 6 | 15 | 25½ | 21 | 16½ | 15 | 13½ |
| 7 | 17½ | 29¾ | 24½ | 19¼ | 17½ | 15¾ |
| 8 | 20 | 34 | 28 | 22 | 20 | 18 |
| 9 | 22½ | 38¼ | 31½ | 24¾ | 22½ | 20¼ |
| 10 | 25 | 42½ | 35 | 27½ | 25 | 22½ |
| 11 | 27½ | 46¾ | 38½ | 30¼ | 27½ | 24¾ |
| 12 | 30 | 51 | 42 | 33 | 30 | 27 |
| 13 | 32½ | 55¼ | 45½ | 35¾ | 32½ | 29¼ |
| 14 | 35 | 59½ | 49 | 38½ | 35 | 31½ |
| 15 | 37½ | 63¾ | 52½ | 41¼ | 37½ | 33¾ |
| 16 | 40 | 68 | 56 | 44 | 40 | 36 |
| 17 | 42½ | 72¼ | 59½ | 46¾ | 42½ | 38¾ |
| 18 | 45 | 76½ | 63 | 49½ | 45 | 40½ |
| 19 | 47½ | 80¾ | 66½ | 52¼ | 47½ | 42¾ |
| 20 | 50 | 85 | 70 | 55 | 50 | 45 |
| 21 | 52½ | 89¼ | 73½ | 57¾ | 52½ | 47¼ |
| 22 | 55 | 93½ | 77 | 60½ | 55 | 49½ |
| 23 | 57½ | 97¾ | 80½ | 63¼ | 57½ | 51¾ |
| 24 | 60 | 102 | 84 | 66 | 60 | 54 |
| 25 | 62½ | 106¼ | 87½ | 68¾ | 62½ | 56¼ |

NOTE—These data apply to obsolete radiators manufactured since 1926 but not manufactured today.

bucket has been replaced by condensate, the bucket loses its buoyancy and when it *sinks*, it opens a valve to permit condensate to be pushed into the return.

*Bucket Trap (Open)*—The bucket (float) is open at the top. Water surrounding the bucket keeps it floating, and the pin is pressed against its seat. Condensate from the system drains into the bucket. When enough has drained into it so that the bucket loses its buoyancy, it sinks and pulls the pin off its seat and steam pressure forces the condensate out of the trap.

*Calorie (Small)*—The quantity of heat required to raise 1 gram of water 1°C (approx.).

*Calorie (Large)*—The quantity of heat required to raise 1 kilogram of water 1°C (approx.).

*Celsius*—A thermometer scale at which the freezing point of water is 0° and its boiling point is 100°.

*Central Fan System*—A mechanical, indirect system of heating, ventilating, or air conditioning consisting of a central plant where

### Table 5-11. Heating Surface—Five-Tube Radiators for Steam or Hot Water

| No. of Sections | Length Inches | HEATING SURFACE—SQUARE FEET 240 Btu PER SQ. FT. PER HOUR | | | | |
|---|---|---|---|---|---|---|
| | | 37" HEIGHT 5 Sq. Ft. Per Sec. | 32" HEIGHT 4⅓ Sq. Ft. Per Sec. | 26" HEIGHT 3½ Sq. Ft. Per Sec. | 23" HEIGHT 3 Sq. Ft. Per Sec. | 20" HEIGHT 2⅔ Sq. Ft. Per Sec. |
| 2 | 5 | 10 | 8⅔ | 7 | 6 | 5⅓ |
| 3 | 7½ | 15 | 13 | 10½ | 9 | 8 |
| 4 | 10 | 20 | 17½ | 14 | 12 | 10⅔ |
| 5 | 12⅓ | 25 | 21⅔ | 17½ | 15 | 13½ |
| 6 | 15 | 30 | 26 | 21 | 18 | 16 |
| 7 | 17½ | 35 | 30⅓ | 24½ | 21 | 18⅔ |
| 8 | 20 | 40 | 34⅔ | 28 | 24 | 21⅓ |
| 9 | 22½ | 45 | 39 | 31½ | 27 | 24 |
| 10 | 25 | 50 | 43⅓ | 35 | 30 | 26⅔ |
| 11 | 27½ | 55 | 47⅔ | 38½ | 33 | 29⅓ |
| 12 | 30 | 60 | 52 | 42 | 36 | 32 |
| 13 | 32½ | 65 | 56½ | 45½ | 39 | 34⅔ |
| 14 | 35 | 70 | 60⅔ | 49 | 42 | 37⅓ |
| 15 | 37½ | 75 | 65 | 52½ | 45 | 40 |
| 16 | 40 | 80 | 69⅓ | 56 | 48 | 42⅔ |
| 17 | 42½ | 85 | 73⅔ | 59½ | 51 | 45⅓ |
| 18 | 45 | 90 | 78 | 63 | 54 | 48 |
| 19 | 47½ | 95 | 82⅓ | 66½ | 57 | 50⅔ |
| 20 | 50 | 100 | 86⅔ | 70 | 60 | 53⅓ |
| 21 | 52½ | 105 | 91 | 73½ | 63 | 56 |
| 22 | 55 | 110 | 05⅓ | 77 | 66 | 58⅔ |
| 23 | 57½ | 115 | 99⅔ | 80½ | 69 | 61⅓ |
| 24 | 60 | 120 | 104 | 84 | 72 | 64 |
| 25 | 62½ | 125 | 108⅓ | 87½ | 75 | 66⅔ |

NOTE—These data apply to obsolete radiators manufactured since 1926 but not manufactured today.

**Courtesy Hoffman Specialty ITT**

### Table 5-12. Heating Surface—Six-Tube Radiators for Steam or Hot Water

| No. of Sections | Length Inches | HEATING SURFACE—SQUARE FEET 240 Btu PER SQ. FT. PER HOUR | | | | |
|---|---|---|---|---|---|---|
| | | 37" HEIGHT 6 Sq. Ft. Per Sec. | 32" HEIGHT 5 Sq. Ft. Per Sec. | 26" HEIGHT 4 Sq. Ft. Per Sec. | 23" HEIGHT 3½ Sq. Ft. Per Sec. | 20" HEIGHT 3 Sq. Ft. Per Sec |
| 2 | 5 | 12 | 10 | 8 | 7 | 6 |
| 3 | 7½ | 18 | 15 | 12 | 10½ | 9 |
| 4 | 10 | 24 | 20 | 16 | 14 | 12 |
| 5 | 12½ | 30 | 25 | 20 | 17½ | 15 |
| 6 | 15 | 36 | 30 | 24 | 21 | 18 |
| 7 | 17½ | 42 | 35 | 28 | 24½ | 21 |
| 8 | 20 | 48 | 40 | 32 | 28 | 24 |
| 9 | 22½ | 54 | 45 | 36 | 31½ | 27 |
| 10 | 25 | 60 | 50 | 40 | 35 | 30 |
| 11 | 27½ | 66 | 55 | 44 | 38½ | 33 |
| 12 | 30 | 72 | 60 | 48 | 42 | 36 |
| 13 | 32½ | 78 | 65 | 52 | 45½ | 39 |
| 14 | 35 | 84 | 70 | 56 | 49 | 42 |
| 15 | 37½ | 90 | 75 | 60 | 52½ | 45 |
| 16 | 40 | 96 | 80 | 64 | 56 | 48 |
| 17 | 42½ | 102 | 85 | 68 | 59½ | 51 |
| 18 | 45 | 108 | 90 | 72 | 63 | 54 |
| 19 | 47½ | 114 | 95 | 76 | 66½ | 57 |
| 20 | 50 | 120 | 100 | 80 | 70 | 60 |
| 21 | 52½ | 126 | 105 | 84 | 73½ | 63 |
| 22 | 55 | 132 | 110 | 88 | 77 | 66 |
| 23 | 57½ | 138 | 115 | 92 | 80½ | 69 |
| 24 | 60 | 144 | 120 | 96 | 84 | 72 |
| 25 | 62½ | 150 | 125 | 100 | 87½ | 75 |

NOTE—These data apply to obsolete radiators manufactured since 1926 but not manufactured today.

Courtesy Hoffman Specialty ITT

the air is heated and/or conditioned and then circulated by fans or blowers through a system of distributing ducts.

*Chimney Effect*—The tendency in a duct or other vertical air passage for air to rise when heated due to its decrease in density.

*Circulating Pipe (Hot-Water System)*—The pipe and orifice in a Hoffman Panelmatic Hot Water Control System through which the return water bypasses the boiler until the temperature of the circulating stream is too low, at which time part of it is replaced by the correct quantity of hot boiler water to restore its temperature.

*Coefficient of Heat Transmission (Overall)-U*—The amount of heat (Btu) transmitted *from air to air* in one hour per square foot of the wall, floor, roof, or ceiling for a difference in temperature of one degree Fahrenheit *between the air on the inside and outside of the wall, floor, roof, or ceiling.*

*Column Radiator*—A type of direct radiator. This radiator has not been listed by manufacturers since 1926.

*Comfort Line*—The effective temperature at which the largest percentage of adults feel comfortable.

### Table 5-13. Heating Surface—Seven-Tube Radiators for Steam or Hot Water

| No. of Sections | Length Inches | HEATING SURFACE—SQUARE FEET 240 Btu PER SQ. FT. PER HOUR | | |
|---|---|---|---|---|
| | | 20" HEIGHT 4¼ Sq. Ft. Per Sec. | 16½" HEIGHT 3½ Sq. Ft. Per Sec. | 13" HEIGHT 2¾ Sq. Ft. Per Sec. |
| 2 | 5 | 8½ | 7 | 5½ |
| 3 | 7½ | 12¾ | 10½ | 8¼ |
| 4 | 10 | 17 | 14 | 11 |
| 5 | 12½ | 21¼ | 17½ | 13¾ |
| 6 | 15 | 25½ | 21 | 16½ |
| 7 | 17½ | 29¾ | 24½ | 19¼ |
| 8 | 20 | 34 | 28 | 22 |
| 9 | 22½ | 38¼ | 31½ | 24¾ |
| 10 | 25 | 42½ | 35 | 27½ |
| 11 | 27½ | 46¾ | 38½ | 30¼ |
| 12 | 30 | 51 | 42 | 33 |
| 13 | 32½ | 55¼ | 45½ | 35¾ |
| 14 | 35 | 59½ | 49 | 38½ |
| 15 | 37½ | 63¾ | 52½ | 41¼ |
| 16 | 40 | 68 | 56 | 44 |
| 17 | 42½ | 72¼ | 59½ | 46¾ |
| 18 | 45 | 76½ | 63 | 49½ |
| 19 | 47½ | 80¾ | 66½ | 52¼ |
| 20 | 50 | 85 | 70 | 55 |
| 21 | 52½ | 89¼ | 73½ | 57¾ |
| 22 | 55 | 93½ | 77 | 60½ |
| 23 | 57½ | 97¾ | 80½ | 63¼ |
| 24 | 60 | 102 | 84 | 66 |
| 25 | 62½ | 106¼ | 87½ | 68¾ |

NOTE—These data apply to obsolete radiators manufactured since 1926 but not manufactured today.

**Courtesy Hoffman Specialty ITT**

*Comfort Zone (Average)*—The range of effective temperatures over which the majority of adults feel comfortable.

*Concealed Radiator*—See Convector.

*Condensate*—In steam heating, the water formed by cooling steam as in a radiator. The capacity of traps, pumps, and so forth, is sometimes expressed in lbs. of condensate they will handle per hour. One lb. of condensate per hour is equal to approximately 4 sq. ft. of steam heating surface (240 Btu/hr per sq. ft.).

*Conductance (Thermal)-C*—The amount of heat (Btu) trans-

**Table 5-14. Heat Emission Standard Column Radiation\***

| Number of Btu transmitted per hour per sq. ft. of radiation with low pressure steam when heating room to given temperature. | | | | | |
|---|---|---|---|---|---|
| Temperature of Room | TYPE OF RADIATION | | | | |
| | 3 Col. 26" | 3 Col. 32" | 3 Col. 38" | Wall. | Coil. |
| Col. A | Col. B | Col. C | Col. D | Col. E | Col. F |
| 40°F. | 309 | 305 | 293 | 362 | 381 |
| 45 | 301 | 292 | 281 | 347 | 365 |
| 50 | 290 | 282 | 271 | 335 | 354 |
| 55 | 279 | 270 | 261 | 322 | 338 |
| 60 | 269 | 261 | 250 | 310 | 326 |
| 65 | 258 | 250 | 240 | 297 | 313 |
| 70 | 247 | 240 | 231 | 285 | 300 |
| 75 | 236 | 230 | 220 | 275 | 288 |
| 80 | 226 | 220 | 211 | 261 | 277 |
| 85 | 216 | 210 | 200 | 251 | 265 |
| 90 | 206 | 200 | 190 | 242 | 253 |
| 95 | 196 | 190 | 180 | 228 | 239 |
| 100 | 186 | 180 | 170 | 215 | 226 |
| 105 | 176 | 171 | 162 | 203 | 214 |
| 110 | 167 | 162 | 155 | 192 | 202 |
| 115 | 158 | 153 | 147 | 181 | 191 |
| 120 | 149 | 144 | 139 | 171 | 180 |
| 125 | 140 | 135 | 130 | 160 | 169 |
| 130 | 130 | 126 | 121 | 150 | 158 |
| 135 | 121 | 118 | 113 | 140 | 147 |
| 140 | 113 | 110 | 106 | 130 | 137 |

If 500 φ of wall radiation is heating a room to 50°F, what will its equivalent be in 32" column radiation setting in 70°F—Refer to Col. E and find one φ wall radiation setting in 50°F = 335 Btu. Then 500 × 335 = 167,500 Btu. Refer to Col. C at 70°F for average Col. Rad. and find 240 Btu per sq. ft. Therefore 167,500 ÷ 240 = 698 φ direct equivalent.

\*This table covers heat emission from radiation manufactured prior to 1926.

**Courtesy Hoffman Specialty ITT**

mitted, *from surface to surface*, in one hour through one square foot of a material or construction *for the thickness or type under consideration* for a difference in temperature of 1°F between the two surfaces.

### Table 5-15. Heat Emission of Direct Pipe Coils Using Steam at 215°F and Room Temp. 70°

Wall Coils—Coils Placed Vertical—Pipes Horizontal
Btu per linear foot of coil per hour (not lineal feet of pipe)

| Size of Pipe Coil Conductor | 1″ | 1¼″ | 1½″ |
|---|---|---|---|
| Single Row | 132 | 162 | 185 |
| Two Row | 252 | 312 | 348 |
| Four Row | 440 | 545 | 616 |
| Six Row | 567 | 702 | 793 |
| Eight Row | 651 | 796 | 907 |
| Ten Row | 732 | 907 | 1020 |
| Twelve Row | 812 | 1005 | 1135 |

Ceiling Coils—Coils placed horizontally—pipes horizontal, emission is equal to that of a single row coil. Allowance must be made, however, if the coil is at the ceiling in a higher temperature. In this case use:

126 Btu per linear foot of pipe for 1″ coils
156 Btu per linear foot of pipe for 1¼″ coils
175 Btu per linear foot of pipe for 1½″ coils

| PIPE SIZE | MEAN WATER °F. | HEAT EMISSION OF HOT WATER COILS (INSTALLATION LIKE STEAM COILS ABOVE) ROWS IN COIL | | | | | | |
|---|---|---|---|---|---|---|---|---|
| | | 1 | 2 | 4 | 6 | 8 | 10 | 12 |
| 1″ | 170 | 80 | 155 | 270 | 350 | 400 | 450 | 500 |
| | 180 | 90 | 175 | 305 | 395 | 450 | 510 | 565 |
| | 190 | 105 | 195 | 345 | 445 | 510 | 570 | 635 |
| | 200 | 115 | 220 | 385 | 495 | 565 | 635 | 705 |
| 1¼″ | 170 | 100 | 190 | 335 | 435 | 490 | 560 | 620 |
| | 180 | 110 | 215 | 380 | 485 | 555 | 630 | 700 |
| | 190 | 125 | 245 | 425 | 550 | 620 | 710 | 785 |
| | 200 | 140 | 270 | 475 | 610 | 690 | 790 | 875 |
| 1½″ | 170 | 115 | 215 | 380 | 490 | 560 | 630 | 700 |
| | 180 | 130 | 240 | 430 | 550 | 630 | 710 | 790 |
| | 190 | 145 | 270 | 480 | 620 | 710 | 795 | 885 |
| | 200 | 160 | 305 | 535 | 690 | 790 | 885 | 985 |

Courtesy Hoffman Specialty ITT

## Table 5-16. Radiator Enclosures

To enclose or partly enclose a radiator reduces its heat output and changes the distribution of heated air in the room. The additional surface usually added to column or tube radiation for various enclosures is indicated below.

*If A is 50% of width of radiator, add 10%; if 150%, add 35%.

‡B = 80% of A.   C = 150% of A.   D = A.

   Example: A room requires 50 sq. ft. radiation radiator recessed flush with wall, −50 φ + 20% − 60 φ radiator required. If radiator for same room is to have grille over entire face only, −50 φ + 30% − 65 φ required.

DEDUCT 10%          ADD 20%          DEDUCT 5%

‡NO CHANGE          ADD 30%          ADD 5%

Courtesy Hoffman Specialty ITT

## Table 5-17. Kitchen Equipment—Pounds of Steam Required Per Hour

|  | Pounds Per Hour |
|---|---|
| Bain Marie—Per Foot | 12.5 |
| Coffee Urn (3) | 50 |
| Egg Boiler—3 Compartment | 10 |
| Stock Kettle—40 Gallon | 37.5 |
| Steam Table—Each 6 Feet | 25 |
| Clam and Lobster Steamer | 10 |
| Jet for Pot Sink | 10 |
| Vegetable Boiler—Per Section | 15 |
| Silver Burnisher | 50 |
| Plate Warmer—15 feet long | 30 |
| Steamer—3 Compartment | 75 |
| Usual Pressure Carried: 30-35 lbs. | |

*Courtesy Hoffman Specialty ITT*

## Table 5-18. Outdoor Storage Tanks (to Keep from Freezing)

$T^1$ = Outdoor Temperature.
$T^2$ = Temperature of Water.
E.S. = Sq. Ft. of Exposed Surface of Tank
Coef. = 2″ Wood Tank—.5 Btu per sq. ft. per hour
Steel Tank—1.5 Btu per sq. ft. per hour
Btu loss from Water per hour = E.S. $\times$ ($T^1$ − $T^2$) $\times$ Coef.

Pounds Steam Required = $\dfrac{\text{Total Btu Loss from Water per Hour}}{\text{Latent Heat Steam at Pressure Carried}}$

Sq. ft. Pipe Coil Required = $\dfrac{\text{Lbs. of Steam required} \div \text{Lbs. Steam}}{\text{Condensed per Hour per sq. ft. of Pipe}}$ (see Table 30)

To obtain sq. ft. of exposed surface for round tanks multiply diameter by 3.1416 $\times$ Height + Area if Bottom and Top.

NOTE—The above is only to be used for water storage tanks, usually located on roof.

*Courtesy Hoffman Specialty ITT*

*Conduction (Thermal)*—The transmission of heat through and by means of matter.

*Conductivity (Thermal)-k*—The amount of heat (Btu) transmitted in one hour through one square foot of a homogeneous mate-

179

rial 1″ thick for a difference in temperature of 1°F between the two surfaces of the material.

*Conductor (Thermal)*—A material capable of readily transmitting heat by means of conduction.

*Convection*—The transmission of heat by the circulation (either

### Table 5-19. Heating Power of Low-Pressure Steam Pipes in Water for Average Working Conditions

| BRASS PIPE | | | | | |
|---|---|---|---|---|---|
| COL. A | COL. B | COL. C | COL. A | COL. B | COL. C |
| Temp. Dif. | Btu | Pounds | Temp. Dif. | Btu | Pounds |
| 6°F. | 192 | .20 | 70 | 13,000 | 13.55 |
| 7 | 240 | .25 | 75 | 15,000 | 15.62 |
| 8 | 300 | .31 | 80 | 17,000 | 17.70 |
| 9 | 400 | .42 | 85 | 19,000 | 20.00 |
| 10 | 480 | .50 | 90 | 21,000 | 21.77 |
| 15 | 800 | .83 | 95 | 23,000 | 24.00 |
| 16 | 960 | 1.00 | 100 | 25,000 | 26.05 |
| 20 | 1,440 | 1.00 | 110 | 30,000 | 31.25 |
| 25 | 2,300 | 2.40 | 120 | 35,000 | 36.45 |
| 30 | 3,100 | 3.23 | 130 | 40,000 | 41.60 |
| 35 | 4,000 | 4.16 | 140 | 45,000 | 46.90 |
| 40 | 5,000 | 5.23 | 150 | 50,000 | 52.10 |
| 45 | 6,000 | 6.25 | 160 | 55,000 | 57.30 |
| 50 | 7,200 | 7.50 | 170 | 61,000 | 63.54 |
| 55 | 8,500 | 8.85 | 180 | 67,000 | 70.00 |
| 60 | 10,000 | 10.50 | 190 | 73,500 | 76.60 |
| 65 | 11,500 | 12.00 | 200 | 80,000 | 83.23 |

Col. A = temperature difference between steam in pipe and average temperature of the water in the tank in degrees.

Col. B = Btu transmitted per sq. ft. per hour.

Col. C = lbs. of steam condensed per sq. ft. per hour. Iron pipe will condense ½ as much steam as given in table for brass pipe.

Example: Heat 200 gals. water per hour from 50° to 164°F with brass pipe and steam at 5 lbs. gauge pressure.

| | |
|---|---|
| Temperature of steam 5 lbs. pressure | = 227°F |
| Average temperature of water (50 + 164) ÷ 2 | = 107°F |
| Temperature difference | = 120°F |
| 200 gals. water by weight (200 × 8⅓) | = 1667 lbs. |
| Temperature rise = 164° − 50° | = 114°F |

Btu required per hr. = 1667 × 114 = 190.038. From Col. A—120°F—find in Col. B that 1 sq. ft. brass pipe will give up 35,000 Btu and in Col. C this equals 36.45 lbs. of steam per hour. Then: 190,038 ÷ 35,000 = 5.43 sq. ft. brass pipe and as each sq. ft. will condense 36.45 lbs.—36.45 × 5.43 = 198 lbs. steam required per hour.

**Courtesy Hoffman Specialty ITT**

natural or forced) of a liquid or a gas such as air. If natural, it is caused by the difference in weight of hotter and colder fluid.

*Convector*—A concealed *radiator*. An enclosed heating unit located (with enclosure) either within, adjacent to, or exterior to the room or space to be heated, but transferring heat to the room or space mainly by the process of convection. A shielded heating unit is also termed a convector. If the heating unit is located exterior to the room or space to be heated, the heat is transferred through one or more ducts or pipes.

*Convertor*—A piece of equipment for heating water with steam

### Table 5-20. Water Flow in G.P.M. Through Pipes for Various Pressure Drops (Not for Heating Systems)

| Pressure Drop Lbs. per Sq. In. per 100 Ft. Run | IRON PIPE SIZE IN INCHES | | | | | | | | |
|---|---|---|---|---|---|---|---|---|---|
| | ³/₄″ | 1″ | 1¹/₄″ | 1¹/₂″ | 2″ | 2¹/₂″ | 3″ | 3¹/₂″ | 4″ |
| 5 | 5.4 | 11 | 19 | 30 | 62 | 109 | 171 | 252 | 353 |
| 7 | 6.4 | 13 | 23 | 36 | 74 | 129 | 203 | 298 | 418 |
| 10 | 7.6 | 15 | 27 | 43 | 88 | 154 | 242 | 357 | 499 |
| 20 | 10.8 | 22 | 38 | 61 | 125 | 218 | 343 | 504 | 706 |
| 30 | 13.2 | 27 | 47 | 76 | 153 | 267 | 420 | 618 | 864 |
| 40 | 15.0 | 31 | 54 | 86 | 176 | 308 | 485 | 714 | 998 |
| 50 | 17.0 | 35 | 60 | 96 | 197 | 345 | 542 | 800 | 1115 |
| 75 | 21.0 | 43 | 74 | 117 | 242 | 423 | 665 | 978 | 1365 |
| 100 | 24.0 | 49 | 85 | 136 | 278 | 485 | 769 | 1130 | 1578 |
| 125 | 27.0 | 55 | 96 | 152 | 311 | 544 | 858 | 1260 | 1765 |
| 150 | 30.0 | 60 | 105 | 166 | 341 | 598 | 939 | 1380 | 1930 |

### Water Flow in G.P.M. Through Type L Copper Tubing for Various Pressure Drops

| Pressure Drop Lbs. per Sq. In. per 100 Ft. Run | TUBING SIZE IN INCHES | | | | | | |
|---|---|---|---|---|---|---|---|
| | ³/₈″ | ¹/₂″ | ³/₄″ | 1″ | 1¹/₄″ | 1¹/₂″ | 2″ |
| 5 | 1.2 | 2.4 | 6.2 | 14. | 23. | 37. | 78. |
| 7 | 1.5 | 2.9 | 7.6 | 16. | 28. | 45. | 93. |
| 10 | 1.8 | 3.5 | 9.4 | 19. | 34. | 53. | 115. |
| 20 | 2.8 | 5.2 | 14.0 | 29. | 50. | 80. | 170. |
| 30 | 3.5 | 6.5 | 18.0 | 37. | 63. | 100. | 215. |
| 40 | 4.1 | 7.8 | 21.0 | 43. | 75. | 120. | 250. |
| 50 | 4.7 | 8.9 | 24.0 | 49. | 85. | 135. | 290. |
| 75 | 5.9 | 11.0 | 30.0 | 61. | 105. | 170. | 360. |
| 100 | 7.0 | 13.0 | 36.0 | 72. | 125. | 200. | 425. |

Courtesy Hoffman Specialty ITT

181

without mixing the two. It may be used for supplying hot water for domestic purposes or for a hot-water heating system.

*Cooling Leg*—A length of uninsulated pipe through which the condensate flows to a trap and which has sufficient cooling surface to permit the condensate to dissipate enough heat to prevent flashing when the trap opens. In the case of a thermostatic trap a cooling leg may be necessary to permit the condensate to drop a sufficient amount in temperature to permit the trap to open.

*Degree Day* (Standard)—A unit that is the difference between 65°F and the daily average temperature when the latter is below 65°F. The "degree days" in any one day is equal to the number of degrees F that the average temperature for that day is below 65°F.

*Dew Point Temperature*—The air temperature corresponding to saturation (100% relative humidity) for a given moisture content. It is the lowest temperature at which air can retain the water vapor it contains.

*Direct-Indirect Heating Unit*—A heating unit located in the room or space to be heated and partially enclosed, the enclosed portion being used to heat air which enters from outside the room.

*Direct Radiator*—Same as Radiator.

*Direct-Return System (Hot Water)*—A two-pipe hot-water system in which the water, after it has passed through a heating unit, is returned to the boiler along a direct path so that the total distance

**Table 5-21. Barometric Pressure in Pounds per Square Inch**

| Barometer Inches | Pressure, psi |
|---|---|
| 28.00 | 13.75 |
| 28.25 | 13.87 |
| 28.50 | 13.99 |
| 28.75 | 14.12 |
| 29.00 | 14.24 |
| 29.25 | 14.36 |
| 29.50 | 14.48 |
| 29.75 | 14.61 |
| 30.00 | 14.73 |
| 30.25 | 14.85 |
| 30.50 | 14.98 |
| 30.75 | 15.10 |
| 31.00 | 15.22 |
| 31.25 | 15.34 |

Courtesy Hoffman Specialty ITT

**Table 5-22. Suction Lift of Pumps with Barometric Pressure at Different Altitudes and Equivalent Head of Water (in Feet)**

| Altitude | Barometric Pressure | Equivalent Head of Water in Feet | Practical Suction Lift |
|---|---|---|---|
| Sea Level | 14.70 lbs sq. in. | 33.95 | 22 ft. |
| ¼ Mile Above | 14.02 lbs sq. in. | 32.38 | 21 ft. |
| ½ Mile Above | 13.33 lbs sq. in. | 30.79 | 20 ft. |
| ¾ Mile Above | 12.66 lbs sq. in. | 29.24 | 18 ft. |
| 1 Mile Above | 12.02 lbs sq. in. | 27.76 | 17 ft. |
| 1¼ Miles Above | 11.42 lbs sq. in. | 26.38 | 16 ft. |
| 1½ Miles Above | 10.88 lbs sq. in. | 25.13 | 15 ft. |
| 2 Miles Above | 9.88 lbs sq. in. | 22.82 | 14 ft. |

**Highest Temp. °F of Condensate Permissible with Vacuum Pumps**

| Vacuum Inches of Mercury In Pump Receiver | Highest Permissible Temperature of Condensate °F |
|---|---|
| 15 | 179 |
| 12 | 187 |
| 10 | 192 |
| 8 | 196 |
| 6 | 201 |
| 4 | 205 |
| 2 | 209 |
| 1 | 210 |

Courtesy Hoffman Specialty ITT

traveled by the water from each radiator is the shortest feasible. There is, therefore, a considerable difference in the lengths of the several circuits composing the system.

*Domestic Hot Water*—Hot water used for purposes other than house heating, such as for laundering, dishwashing, bathing.

*Down-Feed One-Pipe Riser (Steam)*—A pipe that carries steam downward to the heating units and into which the condensation from the heating units drains.

*Down-Feed System (Steam)*—A steam heating system in which

183

the supply mains are above the level of the heating units which they serve.

*Dry-Bulb Temperature*—The temperature of the air as determined by an ordinary thermometer.

## Table 5-23. Length of Expansion Offsets and Bonds for Proper Expansion of Pipe

| Total Expansion in Inches* | FEET OF PIPE IN OFFSET OR U-BEND FOR DIFFERENT DIAMETERS OF PIPE | | | | | | | | | |
|---|---|---|---|---|---|---|---|---|---|---|
| | 2" | 3" | 4" | 5" | 6" | 8" | 10" | 12" | 14" | 16" |
| 1 | 11 | 13 | 15 | 17 | 19 | 21 | 23 | 25 | 27 | 30 |
| 2 | 15 | 18 | 21 | 23 | 26 | 29 | 32 | 35 | 38 | 42 |
| 3 | 18 | 22 | 26 | 29 | 32 | 36 | 40 | 43 | 48 | 52 |
| 4 | 21 | 26 | 30 | 34 | 37 | 42 | 47 | 50 | 56 | 58 |
| 5 | 24 | 30 | 34 | 38 | 41 | 47 | 53 | 57 | 63 | 65 |
| 6 | 27 | 33 | 37 | 41 | 45 | 52 | 58 | 63 | 69 | 71 |
| 7 | 30 | 36 | 40 | 44 | 48 | 56 | 62 | 68 | 74 | |
| 8 | 32 | 39 | 43 | 47 | 52 | 60 | 66 | 72 | | |

*This column shows the total expansion the offset will take care of without a cold strain. In general these amounts can be increased 40 percent, which can be taken up in cold strain of the pipe on being made up.
The length of pipe in the expansion piece should be the same whether in the form of a single right-angle offset or double offset or U-band.
The lengths of arms figured for 12,000 lb. per square inch tension for wrought iron pipe. If steel pipe is used this is good for 16,000 lb. per square inch so that the arm will take care of ⅓ more expansion.

### Expansion of Pipes in Inches per 100 Feet

| Temperature °F | Cast Iron | Wrought Iron | Steel | Brass or Copper |
|---|---|---|---|---|
| 0 | 0.00 | 0.00 | 0.00 | 0.00 |
| 50 | 0.36 | 0.40 | 0.38 | 0.57 |
| 100 | 0.72 | 0.79 | 0.76 | 1.14 |
| 125 | 0.88 | 0.97 | 0.92 | 1.40 |
| 150 | 1.10 | 1.21 | 1.15 | 1.75 |
| 175 | 1.28 | 1.41 | 1.34 | 2.04 |
| 200 | 1.50 | 1.65 | 1.57 | 2.38 |
| 225 | 1.70 | 1.87 | 1.78 | 2.70 |
| 250 | 1.90 | 2.09 | 1.99 | 3.02 |
| 275 | 2.15 | 2.36 | 2.26 | 3.42 |
| 300 | 2.35 | 2.58 | 2.47 | 3.74 |
| 325 | 2.60 | 2.86 | 2.73 | 4.13 |
| 350 | 2.80 | 3.08 | 2.94 | 4.45 |

**Courtesy Hoffman Specialty ITT**

*Dry Return (Steam)*—A return pipe in a steam heating system that carries both water of condensation and air.

*Dry Saturated Steam*—Saturated steam containing no water in suspension.

*Equivalent Direct Radiation (EDR)*—See Square Foot of Heating Surface.

*Extended Heating Surface*—Heating surface consisting of ribs, fins, or ribs that receive heat by conduction from the prime surface.

*Extended Surface Heating Unit*—A heating unit having a relatively large amount of extended surface, which may be integral with the core containing the heating medium or assembled over such a core, making good thermal contact by pressure, or by being soldered to the core or by both pressure and soldering. (An extended surface heating unit is usually placed within an enclosure and therefore functions as a convector.)

*Fahrenheit*—A thermometer scale at which the freezing point of water is 32 and its boiling point is 212 above zero. It is generally used in this country for expressing temperature.

### Table 5-24. Water Required for Humidification

The approximate rule for calculating the amount of water required per hour to maintain any desired humidity in a room is:

Multiply the difference between the number of grains of moisture per cubic foot of air at the required room temperature and humidity and the number of grains per cubic foot of outside air at the given temperature and humidity by the cubic contents of the room by the number of air changes per hour and divide the result by 7000 (this method disregards the expansion of air when heated.)

For most localities, it is customary to assume the average humidity of outside air as 30-40%.

Example:

Grains of moisture at 70° & 40% humidity = 3.19
Grains of moisture at 0° & 30% humitidy = .17
Grains of moisture to be added per cu. ft. = 3.02

Assuming two air changes per hour in a room containing 8000 cu. ft. we have

$$\frac{3.02 \times 8000 \times 2}{7000} = 6.9 \text{ lbs. of water per hour required.}$$

**Courtesy Hoffman Specialty ITT**

## Table 5-25. Properties of Air

| Temperature °F | DRY AIR | | | |
|---|---|---|---|---|
| | Weight per Cu. Ft. of Dry Air in Lbs. | Ratio to Volume at 70°F | Btu Absorbed per Cu. Ft. of Air per °F | Cu. Ft. of Air Raised 1°F by 1 Btu |
| 0 | .08636 | .8680 | .02080 | 48.08 |
| 10 | .08453 | .8867 | .02039 | 49.05 |
| 20 | .08276 | .9057 | .01998 | 50.05 |
| 30 | .08107 | .9246 | .01957 | 51.10 |
| 40 | .07945 | .9434 | .01919 | 52.11 |
| 50 | .07788 | .9624 | .01881 | 53.17 |
| 60 | .07640 | .9811 | .01846 | 54.18 |
| 70 | .07495 | 1.0000 | .01812 | 55.19 |
| 80 | .07356 | 1.0190 | .01779 | 56.21 |
| 90 | .07222 | 1.0380 | .01747 | 57.25 |
| 100 | .04093 | 1.0570 | .01716 | 58.28 |
| 110 | .06968 | 1.0756 | .01687 | 59.28 |
| 120 | .06848 | 1.0945 | .01659 | 60.28 |
| 130 | .06732 | 1.1133 | .01631 | 61.32 |
| 140 | .06620 | 1.1320 | .01605 | 62.31 |
| 150 | .06510 | 1.1512 | .01578 | 63.37 |
| 160 | .06406 | 1.1700 | .01554 | 64.35 |
| 170 | .06304 | 1.1890 | .01530 | 65.36 |
| 180 | .06205 | 1.2080 | .01506 | 66.40 |

Courtesy Hoffman Specialty ITT

*Flash (Steam)*—The rapid passing into steam of water at a high temperature when the pressure it is under is reduced so that its temperature is above that of its boiling point for the reduced pressure. For example: If hot condensate is discharged by a trap into a low-pressure return or into the atmosphere, a certain percentage of the water will be immediately transformed into steam. It is also called reevaporation.

*Float and Thermostatic Trap*—A float trap with a thermostatic element for permitting the escape of air into the return line.

*Float Trap*—A steam trap operated by a float. When enough condensate has drained (by gravity) into the trap body, the float is lifted, which in turn lifts the pin off its seat and permits the condensate to flow into the return until the float has been sufficiently lowered to close the port. Temperature does not affect the operation of a float trap.

*Furnace*—That part of a boiler or warm-air heating plant in which combustion takes place. Sometimes also the complete heating unit of a warm-air heating system.

## Table 5-25. Properties of Air (Cont'd)

| Temperature °F | SATURATED AIR | | | |
|---|---|---|---|---|
| | Vapor Press. Inches of Mercury | Weight of Vapor per Cu. Ft. in Lbs. | Btu Absorbed per Cu. Ft. of Air per °F | Cu. Ft. of Air Raised 1 °F by 1 Btu |
| 0 | 0.0383 | .000069 | .02082 | 48.04 |
| 10 | 0.0631 | .000111 | .02039 | 49.50 |
| 20 | 0.1030 | .000177 | .01998 | 50.05 |
| 30 | 0.1640 | .000276 | .01955 | 51.15 |
| 40 | 0.2477 | .000409 | .01921 | 52.06 |
| 50 | 0.3625 | .000587 | .01883 | 53.11 |
| 60 | 0.5220 | .000829 | .01852 | 54.00 |
| 70 | 0.7390 | .001152 | .01811 | 55.22 |
| 80 | 1.0290 | .001576 | .01788 | 55.93 |
| 90 | 1.4170 | .002132 | .01763 | 56.72 |
| 100 | 1.9260 | .002848 | .01737 | 57.57 |
| 110 | 2.5890 | .003763 | .01716 | 58.27 |
| 120 | 3.4380 | .004914 | .01696 | 58.96 |
| 130 | 4.5200 | .006357 | .01681 | 59.50 |
| 140 | 5.8800 | .008140 | .01669 | 59.92 |
| 150 | 7.5700 | .010310 | .01663 | 60.14 |
| 160 | 9.6500 | .012956 | .01664 | 60.10 |
| 170 | 12.2000 | .016140 | .01671 | 59.85 |
| 180 | 15.2900 | .019940 | .01682 | 59.45 |

Courtesy Hoffman Specialty ITT

*Gauge Pressure*—The pressure above that of the atmosphere. It is the pressure indicated on an ordinary pressure gauge. It is expressed as a unit pressure as lbs. per sq. in. gauge (psig).

*Grille*—A perforated covering for an air inlet or outlet usually made of wire screen, pressed steel, cast iron or other material.

*Head*—Unit pressure usually expressed in feet of water or mil-inches of water.

*Heat*—That form of energy into which all other forms may be changed. Heat always flows from a body of higher temperature to a body of lower temperature. See also, Latent Heat, Sensible Heat, Specific Heat, Heat of the Liquid, Total Heat.

*Heat of the Liquid*—The heat (Btu) contained in a liquid due to its temperature. The heat of the liquid for water is zero at 32°F and increases 1 Btu approximately for every degree rise in temperature.

*Heat Unit*—In the foot-pound-second system, the British thermal unit (Btu); in the centimeter-gram-second system, the calorie (cal.).

*Heating Medium*—A substance such as water, steam, or air used

## Table 5-26. Relative Humidity—Percent

| Dry-Bulb Thermometer °F | DIFFERENCE BETWEEN DRY- AND WET-BULB THERMOMETERS | | | | | | | | | | | | | | | | | | | | | | | | | | |
|---|1.|2.|3.|4.|5.|6.|7.|8.|9.|10.|11.|12.|13.|14.|15.|16.|17.|18.|19.|20.|21.|22.|23.|24.|26.|28.|30.|
| 32 | 89 | 79 | 69 | 59 | 49 | 39 | 30 | 20 | 11 | 2 | | | | | | | | | | | | | | | | | |
| 40 | 92 | 83 | 75 | 68 | 60 | 52 | 45 | 37 | 29 | 23 | 15 | 7 | 0 | | | | | | | | | | | | | | |
| 50 | 93 | 87 | 80 | 74 | 67 | 61 | 55 | 49 | 43 | 38 | 32 | 27 | 21 | 16 | 11 | 5 | 0 | | | | | | | | | | |
| 60 | 94 | 89 | 83 | 78 | 73 | 68 | 63 | 58 | 53 | 48 | 43 | 39 | 34 | 30 | 26 | 21 | 17 | 13 | 9 | 5 | 1 | 0 | | | | | |
| 70 | 95 | 90 | 86 | 81 | 77 | 72 | 68 | 64 | 59 | 55 | 51 | 48 | 44 | 40 | 36 | 33 | 29 | 25 | 22 | 19 | 15 | 12 | 9 | 6 | 0 | | |
| 80 | 96 | 91 | 87 | 83 | 79 | 75 | 72 | 68 | 64 | 61 | 57 | 54 | 50 | 47 | 44 | 41 | 38 | 35 | 32 | 29 | 26 | 23 | 20 | 18 | 12 | 7 | |
| 90 | 96 | 92 | 89 | 85 | 81 | 78 | 74 | 71 | 67 | 65 | 61 | 58 | 55 | 52 | 49 | 47 | 44 | 41 | 39 | 36 | 34 | 31 | 29 | 26 | 22 | 17 | 13 |
| 100 | 96 | 93 | 89 | 86 | 83 | 80 | 77 | 73 | 70 | 68 | 65 | 62 | 59 | 56 | 54 | 51 | 49 | 46 | 44 | 41 | 39 | 37 | 35 | 33 | 28 | 24 | 21 |
| 110 | 97 | 93 | 90 | 87 | 84 | 81 | 78 | 75 | 73 | 70 | 67 | 65 | 62 | 60 | 57 | 55 | 52 | 50 | 48 | 46 | 44 | 42 | 40 | 38 | 34 | 30 | 26 |
| 120 | 97 | 94 | 91 | 88 | 85 | 82 | 80 | 77 | 74 | 72 | 69 | 67 | 65 | 62 | 60 | 58 | 55 | 53 | 51 | 49 | 47 | 45 | 43 | 41 | 38 | 34 | 31 |
| 140 | 97 | 95 | 92 | 89 | 87 | 84 | 82 | 79 | 77 | 75 | 73 | 70 | 68 | 66 | 64 | 62 | 60 | 58 | 56 | 54 | 53 | 51 | 49 | 47 | 44 | 41 | 38 |

EXAMPLE:
DRY-BULB THERMOMETER = 70°
WET-BULB THERMOMETER = 59°
DIFFERENCE = 11°
SATURATION = 100%—BAROMETER 30"

FOLLOW THE HORIZONTAL LINE OPPOSITE
70° DRY-BULB TEMP. TO THE VERTICAL LINE
UNDER 11 = RELATIVE HUMIDITY—51%

to convey heat from the boiler, furnace, or other source of heat to the heating units from which the heat is dissipated.

*Heating Surface*—The exterior surface of a heating unit. See also Extended Heating Surface.

*Heating Unit*—Radiators, convectors, baseboards, finned tubing, coils embedded in floor, wall, or ceiling, or any device that transmits the heat from the heating system to the room and its occupants.

*Horsepower*—A unit to indicate the time rate of doing work equal to 550 ft.-lb. per second, or 33,000 ft.-lb. per minute. One horsepower equals 2545 Btu/hr, or 746 watts.

*Hot-Water Heating System*—A heating system in which water is used as the medium by which heat is carried through pipes from the boiler to the heating units.

*Humidistat*—An instrument that controls the relative humidity of the air in a room.

*Humidity*—The water vapor mixed with air.

*Insulation (Thermal)*—Material with a high resistance to heat flow.

*Latent Heat of Evaporation*—The heat (Btu per pound) neces-

### Table 5-27. Grains of Moisture per Cubic Foot of Air at Various Temperatures and Humidities

| TEMPER-ATURE °F | RELATIVE HUMIDITY, PERCENT | | | | | | | | | |
|---|---|---|---|---|---|---|---|---|---|---|
| | 100% | 90% | 80% | 70% | 60% | 50% | 40% | 30% | 20% | 10% |
| 75 | 9.35 | 8.42 | 7.49 | 6.55 | 5.61 | 4.68 | 3.74 | 2.81 | 1.87 | 0.94 |
| 72 | 8.51 | 7.66 | 6.81 | 5.96 | 5.11 | 4.25 | 3.40 | 2.55 | 1.70 | 0.85 |
| 70 | 7.98 | 7.18 | 6.38 | 5.59 | 4.79 | 3.99 | 3.19 | 2.39 | 1.60 | 0.80 |
| 67 | 7.24 | 6.52 | 5.79 | 5.07 | 4.35 | 3.62 | 2.90 | 2.17 | 1.45 | 0.72 |
| 65 | 6.78 | 6.10 | 5.43 | 4.75 | 4.07 | 3.39 | 2.71 | 2.04 | 1.36 | 0.68 |
| 60 | 5.74 | 5.17 | 4.60 | 4.02 | 3.45 | 2.87 | 2.30 | 1.72 | 1.15 | 0.57 |
| 50 | 4.08 | 3.67 | 3.26 | 2.85 | 2.45 | 2.04 | 1.63 | 1.22 | 0.82 | 0.41 |
| 40 | 2.85 | 2.56 | 2.28 | 1.99 | 1.71 | 1.42 | 1.14 | 0.86 | 0.57 | 0.29 |
| 30 | 1.94 | 1.74 | 1.55 | 1.35 | 1.16 | 0.97 | 0.78 | 0.58 | 0.39 | 0.19 |
| 20 | 1.23 | 1.11 | 0.99 | 0.86 | 0.74 | 0.62 | 0.49 | 0.37 | 0.25 | 0.12 |
| 10 | 0.78 | 0.70 | 0.62 | 0.54 | 0.47 | 0.39 | 0.31 | 0.23 | 0.16 | 0.08 |
| 0 | 0.48 | 0.43 | 0.39 | 0.34 | 0.29 | 0.24 | 0.19 | 0.14 | 0.10 | 0.05 |

7000 Grains of moisture = 1 pound of water.

## Table 5-28. Drying Time and Conditions
## for Representative Material

| KIND AND THICKNESS OF MATERIAL | Temp. °F | Drying Time |
|---|---|---|
| Bedding | 150–190 | 4–6 Hours |
| Cereals | 110–150 | 24 Hours |
| Coconut | 145–155 | 30 Min. |
| Coffee | 160–180 | 2½ Hours |
| Cores—Oil Sand, Molding ½"–1" Thick | 300 | 4½ Hours |
| Cores—Oil Sand, Molding | 480 | 10 Hours |
| Black Sand with Goulac Binder about 8 " ⁶⁄₁₀ of Time for Oil Sand Cores 16" | 480 700 | |
| Feathers | 150–180 | |
| Films—Photographic | 90 | |
| Fruits and Vegetables | 140 | 2–6 Hours |
| Furs | 110 | |
| Glue | 70–90 | 2–4 Hours |
| Glue Size on Furniture | 130 | 4 Hours |
| Gut | 150 | |
| Gypsum Wall Board—Start Wet | 350 | 1 Hour |
| Gypsum Wall Board—Finish | 190 | |
| Gypsum Blocks | 350–190 | 8–16 Hours |
| Hair Goods | 150–190 | 1 Hour |
| Hats—Felt | 140–180 | |
| Hops | 120–180 | |
| Hides—Thin Leather | 90 | 2–4 Hours |
| Ink—Printing | 70–300 | |
| Knitted Fabrics | 140–180 | |
| Leather—Thick Sole | 90 | 4–6 Hours |
| Lumber—Green Hardwood | 100–180 | 3–180 Days |
| Lumber—Green Softwood | 160–220 | 2–14 Days |
| Macaroni | 90–110 | |
| Matches | 140–180 | |
| Milk and Other Liquid Foods (Spray Dried) | 250–300 | Instantaneous |

**Courtesy Hoffman Specialty ITT**

## Table 5-28. Drying Time and Conditions for Representative Material (Cont'd)

| | | |
|---|---|---|
| Molds, Green Sand, C.I. Flasks (1 sur. only ex.) 8" T. | 600 | 6 Hours |
| Molds, Green Sand, C.I. Flasks (1 sur. only ex.) 13" T. | 700 | 13 Hours |
| Nuts | 75-140 | 24 Hours |
| Paper—Glued | 130-300 | |
| Paper—Treated | 140-200 | |
| Rubber | 80-90 | 6-12 Hours |
| Sand, Loose 1" deep | 300 | 10-15 Min. |
| Shade Cloth | 240 | 1-2 Hours |
| Soap | 125 | 12 Hours |
| Starch | 180-200 | 1-4 Hours |
| Stock Feed—Mixed | 180-200 | 20-30 Min. |
| Sugar | 150-200 | 20-30 Min. |
| Tannin and Other Chemicals (Spray Dried) | 250-300 | Instantaneous |
| Terracotta (Air Drying in Conditioning Room) | 155-220 | 12-96 Hours |
| Wall Board | 200-250 | 12-24 Hours |

**Courtesy Hoffman Specialty ITT**

sary to change 1 lb. of liquid into vapor without raising its temperature. In round numbers this is equal to 960 Btu per lb. of water.

*Latent Heat of Fusion*—The heat necessary to melt 1 lb. of a solid without raising the temperature of the resulting liquid. The latent heat of fusion of water (melting 1 lb. of ice) is 144 Btu.

*Mechanical Equivalent of Heat*—The mechanical energy equivalent to 1 Btu, which is equal to 778 ft.-lb.

*Mil-Inch*—One one-thousandth of an inch (0.001").

*One-Pipe Supply Riser (Steam)*—A pipe that carries steam to a heating unit and that also carries the condensation from the heating unit. In an up-feed riser steam travels upward and the condensate downward, while in a down-feed, both steam and condensate travel down.

*One-Pipe System (Hot Water)*—A hot-water heating system in which one pipe serves both as a supply main and also as a return main. The heating units have separate supply and return pipes, but both are connected to the same main.

191

## Table 5-29. Approximate Fuel Equivalents for Home Heating*

| Tons of Coal At 55% Efficiency 12,500 Btu per Lb. | Gallons of Oil At 60% Efficiency 140,000 Btu per Gal. | Cubic Feet of Gas At 75% Efficiency 550 Btu per Cu. Ft. |
|---|---|---|
| 1.0 | 164 | 33,300 |
| 6.1 | 1000 | 203,600 |
| 0.030 | 4.9 | 1000 |

### Approximate Number of Cubic Feet for Ton of Coal

| TYPE | SIZE | CU. FT./TON |
|---|---|---|
| ANTHRACITE | EGG | 38 |
|  | STOVE | 39 |
|  | CHESTNUT | 39 |
|  | PEA & SMALLER | 40 |
| BITUMINOUS | EGG | 48 |
|  | NUT | 44 |
|  | LUMP | 45 |
|  | SCREENINGS | 44 |
|  | MINE RUN | 42 |
| COKE | EGG | 74 |
|  | NUT | 71 |
|  | PEA | 66 |
|  | RANGE | 74 |
| PETROLEUM COKE | LUMP | 74 |
|  | SCREENINGS | 44 |
| POCAHONTAS | EGG | 41 |
|  | NUT | 44 |
|  | LUMP | 38 |
|  | STOVE | 46 |
|  | MINE RUN | 37 |

*This table was compiled from information furnished by the following: Commercial Testing & Engineering Co., Chicago; Anthracite Industries, Inc., New York.

Courtesy Hoffman Specialty ITT

*One-Pipe System (Steam)*—A steam heating system consisting of a main circuit in which the steam and condensate flow in the same pipe. There is but one connection to each heating unit, and it must serve as both the supply and the return.

*Overhead System*—Any steam or hot-water system in which the supply main is above the heating units. With a steam system, the return must be below the heating units; with a water system, the return *may* be above the heating units.

192

## Table 5-30. Water Evaporated and Heat Required for Drying

M = PERCENTAGE OF MOISTURE IN MATERIAL TO BE DRIED
Q = LBS. WATER EVAPORATED PER TON (2000 LBS.) OF DRY MATERIAL
H = BRITISH THERMAL UNITS REQUIRED FOR DRYING PER TON OF DRY MATERIAL

| M | Q | H | M | Q | H | M | Q | H |
|---|---|---|---|---|---|---|---|---|
| 1 | 20.2 | 85,624 | 14 | 325.6 | 424,884 | 35 | 1,077 | 1,269,240 |
| 2 | 40.8 | 108,696 | 15 | 352.9 | 458,248 | 40 | 1,333 | 1,555,960 |
| 3 | 61.9 | 130,424 | 16 | 381.0 | 489,720 | 45 | 1,636 | 1,895,320 |
| 4 | 83.3 | 156,296 | 17 | 409.6 | 521,752 | 50 | 2,000 | 2,303,000 |
| 5 | 105.3 | 180,936 | 18 | 439.0 | 554,680 | 55 | 2,444 | 2,800,280 |
| 6 | 127.7 | 206,024 | 19 | 469.1 | 558,392 | 60 | 3,000 | 3,423,000 |
| 7 | 150.5 | 231,560 | 20 | 500.0 | 623,000 | 65 | 3,714 | 4,222,680 |
| 8 | 173.9 | 257,768 | 21 | 531.6 | 658,392 | 70 | 4,667 | 5,290,040 |
| 9 | 197.8 | 284,536 | 22 | 564.1 | 694,792 | 75 | 6,000 | 6,783,000 |
| 10 | 222.2 | 311,864 | 23 | 597.4 | 732,088 | 80 | 8,000 | 9,023,000 |
| 11 | 247.2 | 339,864 | 24 | 631.6 | 770,392 | 85 | 11,333 | 12,755,960 |
| 12 | 272.7 | 368,424 | 25 | 666.7 | 809,704 | 90 | 18,000 | 20,223,000 |
| 13 | 298.9 | 397,768 | 30 | 857.0 | 1,022,840 | 95 | 38,000 | 42,623,000 |

FORMULA:

$$Q = \frac{2000M}{100 - M} = H$$

$$= 1120Q + 63,000$$

The value of H is found on the assumption that the moisture is heated from 62° to 212° and evaporated at that temperature, and that the specific heat of the material is 0.21[2000 × (212 − 62) × 0.21] = 63,000.

193

## Table 5-31. Heating Values of Fuel

All of the values given here are very approximate but sufficiently close for ordinary calculations. If greater accuracy is required, the exact value should be obtained from the producer of the fuel in question. The heating value is also referred to as the "calorific value."

### COAL

| TYPE | Btu per POUND |
|------|---------------|
| ANTHRACITE | 13,000 |
| SEMI ANTHRACITE | 13,700 |
| BITUMINOUS | 12,500 |
| LIGNITE | 7000 |

### OIL

| GRADE | Btu per GALLON |
|-------|----------------|
| NO. 1 | 135,000 |
| NO. 2 | 140,000 |
| NO. 4 | 155,000 |
| NO. 5 | 150,000 |
| NO. 6 | 153,000 |

### GAS

| KIND | Btu per CUBIC FOOT |
|------|--------------------|
| NATURAL | 1000 |
| MANUFACTURED | 550 |
| PROPANE | 2250 |
| BUTANE | 3000 |

Courtesy Hoffman Specialty ITT

*Panel Heating*—A method of heating involving the installation of the heating units (pipe coils) within the wall, floor, or ceiling of the room.

*Panel Radiator*—A heating unit placed on, or flush with, a flat wall surface and intended to function essentially as a radiator. Do not confuse with panel heating system.

*Plenum Chamber*—An air compartment maintained under pressure and connected to one or more distributing ducts.

*Pressure*—Force per unit area such as lb. per sq. inch. See Static, Velocity, and Total Gauge and Absolute Pressures. Unless otherwise qualified, it refers to unit static gauge pressure.

*Pressure-Reducing Valve*—A piece of equipment for changing the pressure of a gas or liquid from a higher to a lower one.

*Prime Surface*—A heating surface having the heating medium on one side and air (or extended surface) on the other.

*Radiant Heating*—A heating system in which the heating is by radiation only. Sometimes applied to Panel Heating System.

*Radiation*—The transmission of heat in a straight line through space.

**Table 5-32. Specific Heat (Average Values 32° to 212° F.)**

| | | | |
|---|---|---|---|
| Air* | .237 | Ice | .505 |
| Air† | .169 | Iron, Cast | .113 |
| Alcohol | .615 | Kerosene | .500 |
| Aluminum | .212 | Lead | .030 |
| Ammonia | 1.098 | Limestone | .217 |
| Ammonia* | .520 | Marble | .206 |
| Ammonia† | .391 | Mercury | .033 |
| Antimony | .052 | Mica | .208 |
| Asbestos | .195 | Nickel | .109 |
| Benzole | .340 | Nitrogen* | .244 |
| Bismuth | .030 | Nitrogen† | .173 |
| Brass | .092 | Oak | .570 |
| Brick | .220 | Olive Oil | .471 |
| Bronze | .104 | Oxygen* | .224 |
| Carbon, Graphite | .126 | Oxygen† | .155 |
| Carbon, Dioxide* | .215 | Osmium | .031 |
| Carbon, Dioxide† | .168 | Paraffin | .589 |
| Carbon, Monoxide* | .243 | Petroleum | .504 |
| Carbon, Monoxide† | .173 | Platinum | .032 |
| Cement, Portland | .271 | Rubber, Hard | .339 |
| Chalk | .220 | Sand | .195 |
| Chloroform (Liquid) | .235 | Selenium | .068 |
| Chloroform (Gas) | .147 | Silicon | .175 |
| Coal | .201 | Silver | .056 |
| Cobalt | .103 | Steam* | .480 |
| Coke | .0203 | Steam† | .350 |
| Concrete | .156 | Stones | .200 |
| Copper | .092 | Steel | .118 |
| Cork | .485 | Sulphuric Acid | .336 |
| Cotton | .362 | Tantalum | .033 |
| Ether | .540 | Tin | .054 |
| Fuel Oil | .500 | Turpentine | .420 |
| Gasoline | .500 | Tungsten | .034 |
| Glass | .180 | Water | 1.000 |
| Gold | .032 | Wool | .393 |
| Gypsum | .259 | Wood | .327 |
| Hydrogen* | 3.41 | Zinc | .093 |
| Hydrogen† | 2.81 | | |

*Constant Pressure.
†Constant Volume.

## Table 5-33. Specific Gravity of Metals

| METAL | SPECIFIC GRAVITY |
|---|---|
| Water (Basis for Comparison) | 1.00 |
| Aluminum | 2.55–2.80 |
| Tin (Cast) | 7.2–7.5 |
| Steel | 7.84 |
| Cast Iron | 7.03–7.13 |
| Wrought Iron | 7.6–7.9 |
| Brass | 8.4–8.7 |
| Copper | 8.8–8.95 |
| Lead (Cast) | 11.35 |
| Mercury | 13.60 |
| Platinum | 21.50 |

Courtesy Hoffman Specialty ITT

*Radiator*—A heating unit located within the room to be heated and exposed to view. A radiator transfers heat by radiation to objects "it can see" and by conduction to the surrounding air, which in turn is circulated by natural convection.

*Recessed Radiator*—A heating unit set back into a wall recess but not enclosed.

*Reducing Valve*—See Pressure-Reducing Valve.

*Reevaporation*—See Flash.

*Refrigeration, Ton of*—See Ton of Refrigeration.

*Register*—A grille with a built-in damper or shutter.

*Relative Humidity*—The amount of moisture in a given quantity of air compared with the maximum amount of moisture the same quantity of air could hold at the same temperature. It is expressed as a percentage.

*Return Mains*—The pipes that return the heating medium from the heating units to the source of heat supply.

*Reverse-Return System (Hot Water)*—A two-pipe hot-water heating system in which the water from the several heating units is returned along paths arranged so that all radiator circuits of the system are practically of equal length.

## Table 5-34. Lineal Expansion of Solids at Ordinary Temperatures

| MATERIAL | For 1 °F, Length = 1" |
|---|---|
| Aluminum (Cast) | 0.00001234 |
| Brass Cast | 0.00000957 |
| Brass Plate | 0.00001052 |
| Brick (Fire) | 0.00000300 |
| Bronze (Copper, 96½; Tin, 2½; Zinc, 1) | 0.00000986 |
| Copper | 0.00000887 |
| Glass, Hard | 0.00000397 |
| Gold, Pure | 0.00000786 |
| Iron, Wrought | 0.00000648 |
| Iron, Cast | 0.00000556 |
| Lead | 0.00001571 |
| Mercury (Cubic Expansion) | 0.00009984 |
| Nickel | 0.00000695 |
| Porcelain | 0.00000200 |
| Silver, Pure | 0.00001079 |
| Slate | 0.00000577 |
| Steel, Cast | 0.00000636 |
| Steel, Tempered | 0.00000689 |
| Stone (Sandstone), Dry | 0.00000652 |
| Tin | 0.00001163 |
| Wood, Pine | 0.00000276 |
| Zinc | 0.00001407 |

Courtesy Hoffman Specialty ITT

*Roof Ventilator*—A device placed on the roof of a building to permit egress of air.

*Sensible Heat*—Heat that only increases the temperature of objects, as opposed to latent heat.

*Specific Heat*—In the foot-pound-second system, the amount of heat (Btu) required to raise 1 lb. of a substance 1°F. In the centimeter-gram second system, the amount of heat (cal.) required to raise one gram of a substance 1°C. The specific heat of water is 1.

197

## Table 5-35. Weights of Materials

| MATERIAL | Pounds per Cubic Ft. | MATERIAL | Pounds per Cubic Ft. |
|---|---|---|---|
| Aluminum | 166.5 | Lead | 709.7 |
| Ashes | 45–50 | Lignite | 31–47 |
| Barley | 37–40 | Lime | 50–80 |
| Brass—Copper, Zinc | | Limestone | 156–162 |
| 80   20 | 536.3 | Manganese | 450 |
| 70   30 | 523.8 | Mercury 32° | 849.3 |
| 60   40 | 521.3 | Mercury 60° | 846.8 |
| 50   50 | 511.4 | Mercury 212° | 834.4 |
| Bronze—Cop. 95 to 80 | 552 | Nickel | 548.7 |
| Tin   5 to 20 | | Oats | 25–30 |
| Cement | 90–118 | Ore (Iron) | 105–215 |
| Charcoal | 17–27 | Platinum | 1333 |
| Clay | 95–169 | Rye | 44–50 |
| Coal (Lump) | 50–54 | Sand | 75–120 |
| Nut Coal & Screenings | 53–60 | Silver | 655.1 |
| Coke | 26–30 | Slag (Blast Furnace) | 37–63 |
| Earth | 75–115 | Steel | 489.6 |
| Gold, Pure | 1200.9 | Stone | 90–120 |
| Copper | 552 | Wheat | 458.3 |
| Gravel | 90–135 | Wheat | 44–50 |
| Iron, Cast | 450 | Zinc | 448 |
| Iron, Wrought | 480 | | |

**Courtesy Hoffman Specialty ITT**

*Split System*—A system in which the heating is accomplished by radiators or convectors and ventilation by separate apparatus.

*Square Foot of Heating Surface*—Equivalent direct radiation (EDR). By definition, that amount of heating surface that will give off 240 Btu/hr when filled with a heating medium at 215°F and surrounded by air at 70°F. The equivalent square foot of heating surface may have no direct relation to the actual surface area.

*Static Pressure*—The pressure that tends to burst a pipe. It is used to overcome the frictional resistance to flow through the pipe. It is expressed as a unit pressure and may be either in absolute or gauge pressure. It is frequently expressed in feet of water column or (in the case of pipe friction) in mil-inches of water column per foot of pipe.

## Table 5-36. Melting Points of Metals

| | °F | | °F |
|---|---|---|---|
| Aluminum | 1220 | Iron (Cast) Gray | 2460–2550 |
| Antimony | 1167 | Iron (Cast) White | 1920–2010 |
| Bismuth | 520 | Iron, Wrought | 2460–2640 |
| Brass (Red) | 1870 | Lead | 622 |
| Bronze | 1900 | Silver (Pure) | 1751 |
| Copper | 1981 | Steel | 2370–2550 |
| Glass | 2377 | Tin | 449 |
| Gold (Pure) | 1945 | Zinc | 787 |
| Solder (Lead-Tin) | 350–570 | | |

**Courtesy Hoffman Specialty ITT**

## Table 5-37. Boiling Points of Various Fluids

| | °F | | °F |
|---|---|---|---|
| Water (Atmospheric Pressure) | 212 | Turpentine | 320 |
| Alcohol | 173 | Sulphur | 832 |
| Sulphuric Acid | 620 | Linseed Oil | 549 |

**Courtesy Hoffman Specialty ITT**

*Steam*—Water in the vapor phase. The vapor formed when water has been heated to its boiling point, corresponding to the pressure it is under. See also Dry Saturated Steam, Wet Saturated Steam, Superheated Steam.

*Steam Heating System*—A heating system in which the heating units give up their heat to the room by condensing the steam furnished to them by a boiler or other source.

*Steam Trap*—A device for allowing the passage of condensate and air but preventing the passage of steam. See Thermostatic, Float, Bucket Trap.

*Superheated Steam*—Steam heated above the temperature corresponding to its pressure.

*Supply Mains*—The pipes through which the heating medium flows from the boiler or source of supply to the runouts and risers leading to the heating units.

*Tank Regulator*—See Temperature Regulator.

*Temperature Regulator*—A piece of equipment for controlling the admission of steam to a hot-water (or other liquid) heating

199

## Table 5-38. Conversion Table
## Metric to English

| METRIC | ENGLISH | |
|---|---|---|
| **MEASURES OF LENGTH** | | |
| 1 Kilometer | 0.621 | Mile |
| 1000 Meters | 3281. | Feet |
| 1 Meter | 1.094 | Yards |
| 100 Centimeters | 3.28 | Feet |
| 1000 Millimeters | 39.37 | Inches |
| 1 Centimeter | 0.0328 | Feet |
| 10 Millimeters | 0.394 | Inches |
| 1 Millimeter | 0.0394 | Inches |
| **MEASURES OF SURFACE** | | |
| 1 sq. Kilometer | 0.386 | Sq. Mile |
| 1,000,000 Sq. Meters | 247.1 | Acres |
| | 1,195,985 | Sq. Yards |
| 1 Sq. Meter | 1.196 | Sq. Yards |
| 10,000 Sq. Centimeters | 10.76 | Sq. Feet |
| | 1550. | Sq. Inches |
| 1 Sq. Centimeter | 0.155 | Sq. Inch |
| 100 Sq. Millimeters | 0.0011 | Sq. Feet |
| 1 Sq. Millimeter | 0.00155 | Sq. Inch |
| **MEASURES OF VOLUME AND CAPACITY** | | |
| 1 Cu. Meter | 1.308 | Cu. Yards |
| 1000 Liters | 35.31 | Cu. Feet |
| 1,000,000 Cu. Centimeters | 61023.4 | Cu. Inches |
| 1 Liter | 0.264 | Gallons (U.S.) |
| 1,000 Cu. Centimeters | 0.220 | Gallons (Imperial) |
| | 1.057 | Quarts (U.S.) |
| | 0.880 | Quarts (Imperial) |
| 1 Cu. Centimeter | 0.061 | Cu. Inches |
| 1000 Cu. Millimeters | | |
| **MEASURES OF WEIGHT** | | |
| 1 Kilogram | 0.0011 | Ton (2000 Lbs.) |
| 1000 Gram | 2.205 | Pounds (Av.) |
| 1 Gram | 0.0022 | Pounds (Av.) |
| | 0.035 | Ounces (Av.) |
| | 15.43 | Grains |

NOTE—For conversion of pressures from metric to English units, see Table 5-39.

Courtesy Hoffman Specialty ITT

## Table 5-39. Conversion Table (Pressures)

| INCHES Water | INCHES Mercury | FEET Water | MILLIMETER Mercury (Hg) | METER Water | POUNDS Per Sq. Inch | POUNDS Per Sq. Ft. | GRAMS Sq. Centimeter | KILOGRAMS Per Sq. Meter | KILOGRAMS Per Sq. Cm. | Atmosphere |
|---|---|---|---|---|---|---|---|---|---|---|
| 1.0 | 0.074 | 0.083 | 1.88 | 0.0254 | 0.036 | 5.20 | 2.53 | 25.37 | 0.003 | 0.0024 |
| 13.6 | 1.0 | 1.13 | 25.4 | 0.344 | 0.490 | 70.5 | 34.4 | 344.4 | 0.0344 | 0.0333 |
| 12.0 | 0.884 | 1.0 | 22.4 | 0.305 | 0.433 | 62.4 | 30.4 | 304.5 | 0.0304 | 0.0295 |
| 0.54 | 0.039 | 0.045 | 1.0 | 0.014 | 0.019 | 2.78 | 1.36 | 13.6 | 0.00136 | 0.00131 |
| 39.4 | 2.89 | 3.28 | 73.5 | 1.0 | 1.422 | 204.6 | 100.0 | 1000.0 | 0.10 | 0.0967 |
| 27.7 | 2.04 | 2.390 | 51.8 | 0.703 | 1.0 | 144.0 | 70.3 | 703.1 | 0.070 | 0.0680 |
| 0.19 | 0.014 | 0.016 | 0.36 | 0.005 | 0.0069 | 1.0 | 0.49 | 4.88 | 0.00049 | 0.00047 |
| 0.40 | 0.03 | 0.033 | 0.74 | 0.01 | 0.014 | 2.05 | 1.0 | 10.0 | 0.001 | 0.00097 |
| 0.04 | 0.003 | 0.0033 | 0.074 | 0.001 | 0.0014 | 0.205 | 0.10 | 1.0 | 0.0001 | 0.0001 |
| 393.8 | 28.96 | 32.8 | 735.5 | 10.0 | 14.2 | 2048. | 1000.0 | 10000.0 | 1.0 | 0.9678 |
| 407. | 29.92 | 33.9 | 760.0 | 10.3 | 14.7 | 2116. | 1033.2 | 10332. | 1.03 | 1.0 |

To use table go to column headed by unit to be converted. Follow this column down to the 1.0 in heavy print and read horizontally across. Example: Convert five kilograms per sq. meter to lbs. per sq. inch. Select column headed kilograms per sq. meter and follow down to 1.0 then to left column headed lbs per sq. inch. The number 0.0014 found in this space is the conversion factor by which the number of kilograms per sq. meter must be multiplied to change this quantity to lbs. per sq. inch. Therefore, 5 kilograms per meter is equal to five times 0.0014, or 0.007 lb. per sq. inch.

Courtesy Hoffman Specialty ITT

device in the correct quantities so that the temperature of the liquid will remain constant.

*Thermostat*—An instrument that responds to changes in temperature and directly or indirectly controls the room temperature.

*Thermostatic Trap*—A steam trap that opens by a drop in temperature, such as when cold condensate or air reaches it, and closes it when steam reaches it. The temperature sensitive element is usually a sealed bellows or series of diaphragm chambers containing a small quantity of volatile liquid.

*Ton of Refrigeration*—The heat that must be extracted from one ton (2000 lbs.) of water at 32°F to change it into ice at 32°F in 24 hours. It is equal to 288,000 Btu/24 hrs, 12,000 Btu/hr, or 200 Btu/min.

*Total Heat*—The latent heat of vaporization added to the heat of the liquid with which it is in contact.

*Total Pressure*—The sum of the static and velocity pressures. It

### Table 5-40. Conversion Table (Inches to Centimeters to Millimeters)

| in. | cm. | mm. |
|---|---|---|
| 1.00 | 2.54 | 25.40 |
| 2.00 | 5.08 | 50.80 |
| 3.00 | 7.62 | 76.20 |
| 4.00 | 10.16 | 101.60 |
| 5.00 | 12.70 | 127.00 |
| 6.00 | 15.24 | 152.40 |
| 7.00 | 17.78 | 177.80 |
| 8.00 | 20.32 | 203.20 |
| 9.00 | 22.86 | 228.60 |
| 10.00 | 25.40 | 254.00 |
| 20.00 | 50.80 | 508.00 |
| 30.00 | 76.20 | 762.00 |
| 36.00 | 91.40 | 914.00 |
| 40.00 | 101.60 | 1016.00 |
| 50.00 | 127.00 | 1270.00 |
| 60.00 | 152.40 | 1524.00 |
| 70.00 | 177.80 | 1778.00 |
| 80.00 | 203.20 | 2032.00 |
| 90.00 | 228.60 | 2286.00 |
| 100.00 | 254.00 | 2540.00 |

## Table 5-41. Fahrenheit-Celsius Conversion Table

| F | C | F | C | F | C | F | C |
|---|---|---|---|---|---|---|---|
| -20 | -28.9 | 62 | 16.7 | 144 | 62.2 | 226 | 107.8 |
| -18 | -27.8 | 64 | 17.8 | 146 | 63.3 | 228 | 108.9 |
| -16 | -26.7 | 66 | 18.9 | 148 | 64.4 | 230 | 110. |
| -14 | -25.6 | 68 | 20 | 150 | 65.6 | 232 | 111.1 |
| -12 | -24.4 | 70 | 21.1 | 152 | 66.7 | 234 | 112.2 |
| -10 | -23.3 | 72 | 22.2 | 154 | 67.8 | 236 | 113.3 |
| -8 | -22.2 | 74 | 23.3 | 156 | 68.9 | 238 | 114.4 |
| -6 | -21.1 | 76 | 24.4 | 158 | 70 | 240 | 115.6 |
| -4 | -20 | 78 | 25.6 | 160 | 71.1 | 242 | 116.7 |
| -2 | -18.9 | 80 | 26.7 | 162 | 72.2 | 244 | 117.8 |
| 0 | -17.8 | 82 | 27.8 | 164 | 73.3 | 246 | 118.9 |
| 2 | -16.7 | 84 | 28.9 | 166 | 74.4 | 248 | 120 |
| 4 | -15.6 | 86 | 30 | 168 | 75.6 | 250 | 121.1 |
| 6 | -14.4 | 88 | 31.1 | 170 | 76.7 | 252 | 122.2 |
| 8 | -13.3 | 90 | 32.2 | 172 | 77.8 | 254 | 123.3 |
| 10 | -12.2 | 92 | 33.3 | 174 | 78.9 | 256 | 124.4 |
| 12 | -11.1 | 94 | 34.4 | 176 | 80 | 258 | 125.6 |
| 14 | -10 | 96 | 35.6 | 178 | 81.1 | 260 | 126.7 |
| 16 | -8.9 | 98 | 36.7 | 180 | 82.2 | 262 | 127.8 |
| 18 | -7.8 | 100 | 37.8 | 182 | 83.3 | 264 | 128.9 |
| 20 | -6.7 | 102 | 38.9 | 184 | 84.4 | 266 | 130. |
| 22 | -5.6 | 104 | 40. | 186 | 85.6 | 268 | 131.1 |
| 24 | -4.4 | 106 | 41.1 | 188 | 86.7 | 270 | 132.2 |
| 26 | -3.3 | 108 | 42.2 | 190 | 87.8 | 272 | 133.3 |
| 28 | -2.2 | 110 | 43.3 | 192 | 88.9 | 274 | 134.4 |
| 30 | -1.1 | 112 | 44.4 | 194 | 90 | 276 | 135.6 |
| 32 | 0. | 114 | 45.6 | 196 | 91.1 | 278 | 136.7 |
| 34 | 1.1 | 116 | 46.7 | 198 | 92.2 | 280 | 137.8 |
| 36 | 2.2 | 118 | 47.8 | 200 | 93.3 | 282 | 138.9 |
| 38 | 3.3 | 120 | 48.9 | 202 | 94.4 | 284 | 140. |
| 40 | 4.4 | 122 | 50. | 204 | 95.6 | 286 | 141.1 |
| 42 | 5.6 | 124 | 51.1 | 206 | 96.7 | 288 | 142.2 |
| 44 | 6.7 | 126 | 52.2 | 208 | 97.8 | 290 | 143.3 |
| 46 | 7.8 | 128 | 53.3 | 210 | 98.9 | 292 | 144.4 |
| 48 | 8.9 | 130 | 54.4 | 212 | 100. | 294 | 145.6 |
| 50 | 10. | 132 | 55.6 | 214 | 101.1 | 296 | 146.7 |
| 52 | 11.1 | 134 | 56.7 | 216 | 102.2 | 298 | 147.8 |
| 54 | 12.2 | 136 | 57.8 | 218 | 103.3 | 300 | 148.9 |
| 56 | 13.3 | 138 | 58.9 | 220 | 104.4 | | |
| 58 | 14.4 | 140 | 60. | 222 | 105.6 | | |
| 60 | 15.6 | 142 | 61.1 | 224 | 106.7 | | |

Courtesy Hoffman Specialty ITT

is also used as the total static pressure over an entire area, that is, the unit pressure multiplied by the area on which it acts.

*Trap*—See Steam Trap, Thermostatic Trap, Float Trap, and Bucket Trap.

*Two-Pipe System (Steam or Water)*—A heating system in which one pipe is used for the supply main and another for the return main. The essential feature of a two-pipe hot-water system is that each heating unit receives a direct supply of the heating medium, which cannot have served a preceding heating unit.

*Unit Heater*—A heating unit consisting of a heat-transfer element, a housing, a fan with driving motor, and outlet deflectors or diffusers. It is usually suspended from the ceiling and its heat

## Table 5-42. Electrical Units

**VOLT**—The unit of electrical motive force; force required to send 1 ampere of current through 1 ohm of resistance.

**OHM**—Unit of resistance. The resistance offered to the passage of 1 ampere, when impelled by 1 volt.

**AMPERE**—Unit of current; the current that 1 volt can send through a resistance of 1 ohm.

**COULOMB**—Unit of quantity. Quantity of current that, impelled by 1 volt, would pass through 1 ohm in 1 second.

**JOULE**—Unit of work. The work done by 1 watt in 1 second.

**WATT**—The unit of electrical energy that is the product of ampere and volt. That is, 1 ampere of current flowing under a pressure of 1 volt gives 1 watt of energy. One electrical horsepower is equal to 746 watts.

**ONE KILOWATT**—Is equal to 1000 watts, or 3415 Btu, when used for heating or the equivalent output of 14.2 sq. ft. of steam radiation.

**ONE KILOWATT HOUR**—(kWh) equals the consumption of 1000 watts in one hour.

To find the watts consumed in a given electrical circuit, multiply the volts by the amperes.

To find the volts—divide the watts by the amperes.

To find the amperes—divide the watts by the volts.

To find the watts consumed in a given electrical circuit, multiply the volts by motor by 746. With ac current multiply the wattage by the power factor, then divide by 74.6.

To find the amperes of a given circuit, of which the volts and ohms resistance are known, divide the volts by the ohms.

To find the volts, when the amperes and ohms are known, multiply the amperes by the ohms.

To find the resistance in ohms, when the volts and amperes are known, divide the volts by the amperes.

## Table 5-43. Capacity of Round Storage Tanks

| Depth or Length | NUMBER OF GALLONS | | | | | | | | | |
|---|---|---|---|---|---|---|---|---|---|---|
| | 18 | 24 | 30 | 36 | 12 | 48 | 54 | 60 | 66 | 72 |
| 1 In. | 1.10 | 1.96 | 3.06 | 4.41 | 5.99 | 7.83 | 9.91 | 12.24 | 14.81 | 17.62 |
| 2 Ft. | 26 | 47 | 73 | 105 | 144 | 188 | 238 | 294 | 356 | 423 |
| 2½ | 33 | 59 | 91 | 131 | 180 | 235 | 298 | 367 | 445 | 530 |
| 3 | 40 | 71 | 100 | 158 | 216 | 282 | 357 | 440 | 534 | 635 |
| 3½ | 46 | 83 | 129 | 184 | 252 | 329 | 416 | 513 | 623 | 740 |
| 4 | 53 | 95 | 147 | 210 | 288 | 376 | 475 | 586 | 712 | 846 |
| 4½ | 59 | 107 | 165 | 238 | 324 | 423 | 534 | 660 | 800 | 952 |
| 5 | 66 | 119 | 181 | 264 | 360 | 470 | 596 | 734 | 899 | 1057 |
| 5½ | 73 | 130 | 201 | 290 | 396 | 517 | 655 | 808 | 979 | 1163 |
| 6 | 79 | 141 | 219 | 315 | 432 | 564 | 714 | 880 | 1066 | 1268 |
| 6½ | 88 | 155 | 236 | 340 | 468 | 611 | 770 | 954 | 1156 | 1374 |
| 7 | 92 | 165 | 255 | 368 | 504 | 658 | 832 | 1028 | 1244 | 1480 |
| 7½ | 99 | 179 | 278 | 396 | 540 | 705 | 889 | 1101 | 1335 | 1586 |
| .8 | 106 | 190 | 291 | 423 | 576 | 752 | 949 | 1175 | 1424 | 1691 |
| 9 | 119 | 212 | 330 | 476 | 648 | 846 | 1071 | 1322 | 1599 | 1903 |
| 10 | 132 | 236 | 366 | 529 | 720 | 940 | 1189 | 1463 | 1780 | 2114 |
| 12 | 157 | 282 | 440 | 634 | 684 | 1128 | 1428 | 1762 | 2133 | 2537 |
| 14 | 185 | 329 | 514 | 740 | 1008 | 1316 | 1666 | 2056 | 2490 | 2960 |
| 16 | 211 | 376 | 587 | 846 | 1152 | 1504 | 1904 | 2350 | 2844 | 3383 |
| 18 | 238 | 423 | 660 | 952 | 1296 | 1692 | 2140 | 2640 | 3200 | 3806 |
| 20 | 264 | 470 | 734 | 1057 | 1440 | 1880 | 2380 | 2932 | 3556 | 4230 |

### RECTANGULAR TANK

To find the capacity in U.S. gallons of rectangular tanks, reduce all dimensions to inches, then mulitiply the length by the width by the height and divide the product by 231.

Example:

Tank 56" long × 32" × wide × 20" deep
Then 56" × 32" × 20" = 35840 cu. in.
35840 ÷ 231 = 155 gals. capacity

output is controlled by starting and stopping the fan by a room thermostat. The circulation of the heating medium (steam or hot water) is usually continuous. It is used mostly for industrial heating.

*Unit Pressure*—Pressure per unit area as lbs. per sq. in. (psi)

*Up-Feed System (Hot Water or Steam)*—A heating system in which the supply mains are below the level of the heating units they serve.

*Vacuum Heating System (Steam)*—A one- or two-pipe heating

## Table 5-44. Useful Data

| | | | |
|---|---:|---:|---|
| Diameter | × | 3.1416 | = Circumference |
| Circumference | × | .3183 | = Diameter |
| Diameter$^2$ | × | .7854 | = Area of Circle |
| Area of Circle | × | 1.2732 | = Area of Circumscribed Square |
| Area of Circle | × | .63662 | = Area of Inscribed Square |
| Diameter of Circle | × | .88623 | = Side of Equal Square |
| Diameter of Circle | × | .7071 | = Side of Inscribed Square |
| Circumference of Circle | × | 1.1284 | = Perimeter of Equal Square |
| Side of Square | × | 1.4142 | = Diameter of Circumscribed Circle |
| Side of Square | × | 1.1284 | = Diameter of Equal Circle |
| Perimeter of Square | × | .88623 | = Circumference of Equal Circle |
| Diameter$^2$ | × | 3.1416 | = Surface of Sphere |
| Diameter | × | .5236 | = Volume of Sphere |
| Diameter of Sphere | × | .806 | = Dimensions of Equal Cube |
| Diameter of Sphere | × | .667 | = Length of Equal Cylinder |
| Area of Base | × | ⅓ Height | = Volume of Pyramid or Cone |
| Base | × | ½ Height | = Area of Triangle |
| Radius | × | 1.1547 | = Side of Inscribed Cube |
| Square Inches | × | 1.2732 | = Circular Inches |
| Square Inches | × | .00695 | = Square Feet |
| Square Feet | × | .111 | = Square Yard |
| Square Yards | × | .0002066 | = Acres |
| Cubic Inches | × | .00058 | = Cubic Feet |
| Cubic Feet | × | .03704 | = Cubic Yards |
| Cubic Inches | × | .004329 | = U.S. Gallons |
| Cubic Feet | × | 7.4805 | = U.S. Gallons |
| Cubic Inches | × | .000466 | = U.S. Bushels |
| Cubic Feet | × | .8036 | = U.S. Bushels |
| U.S. Bushels | × | 2150.42 | = Cubic Inches |
| U.S. Bushels | × | 1.242 | = Cubic Feet |
| U.S. Bushels | × | .046 | = Cubic Yards |
| U.S. Gallons | × | 231.0 | = Cubic Inches |
| U.S. Gallons | × | .13368 | = Cubic Feet |
| Cubic Inches Water | × | .036127 | = Pounds (Avoirdupois) |
| Cubic Feet Water | × | 62.4283 | = Pounds (Avoirdupois) |
| U.S. Gallons Water | ÷ | 268.8 | = Tons |
| Column of Water 1″ Diameter x 12″ High } | | | = .34 Pounds (Avoirdupois) |
| Cubic Inch | × | .263 | = Pound Average Cast Iron |
| Cubic Inch | × | .281 | = Pound Average Wrought Iron |
| Cubic Inch | × | .283 | = Pound Average Cast Steel |
| Cubic Inch | × | .3225 | = Pound Average Copper |
| Cubic Inch | × | .3037 | = Pound Average Brass |
| Cubic Inch | × | .26 | = Pound Average Zinc |
| Cubic Inch | × | .4103 | = Pound Average Lead |
| Cubic Inch | × | .2636 | = Pound Average Tin |
| Cubic Inch | × | .4908 | = Pound Average Mercury |
| 12 × Weight of Pine Pattern | | | = Iron Casting |
| 13 × Weight of Pine Pattern | | | = Brass Casting |
| 14 × Weight of Pine Pattern | | | = Lead Casting |
| 1—Calorie | | | = 3.968 Btu |
| 1—Btu | | | = 0.252 Calorie |
| 1—Pound per Square Inch | | | = 703.08 Kilograms per M$^2$ |
| 1—Kilogram per M$^2$ | | | = 0.00142 Pounds per Square Inch |
| 1—Calorie per M$^2$ | | | = 0.3687 Btu per Square Foot |
| 1—Btu per Square Foot | | | = 2.712 Calories per M$^2$ |
| 1—Calorie per M$^2$ per Degree Difference Centigrade } | | | { = 0.2048 Btu per Square Foot per Degree Difference, Fahrenheit |

## Table 5-44. Useful Data (Cont'd)

| | |
|---|---|
| 1—Btu per Square Foot per Degree Difference Fahrenheit | = 4.882 Calories per $M^2$ per Degree Difference Fahrenheit |
| 1—Btu per Pound | = 0.556 Calorie per Kilogram |
| 1—Calorie per Kilogram | = 1.8 Btu per Pound |
| 1—Liter of Coke at 26.3 Lb. per Cu. Ft. | = 0.93 Pounds |
| 1—Lb. of Coke at 26.3 per Cu. Ft. | = 1.076 Liter |
| Water Expands in Bulk from 40° to 212° | = One Twenty-Third |

Courtesy Hoffman Specialty ITT

system equipped with the necessary accessory apparatus to permit the pressure in the system to go below atmospheric.

*Vapor*—Any substance in the gaseous state.

*Vapor Heating System (Steam)*—A two-pipe heating system that operates under pressure at or near atmospheric and returns the condensation to the boiler or receiver by gravity.

*Velocity Pressure*—The pressure used to create the velocity of flow in a pipe. It is expressed as a unit pressure.

*Ventilation*—Air circulated through a room for ventilating pur-

## Table 5-45. Weights of Steel and Iron Plates and Sheets (U.S. Standard Gauge)

| No. of Gauge | Thickness Inches (Approx.) | WEIGHT PER SQUARE FOOT | |
|---|---|---|---|
| | | Iron | Steel |
| 0000000 | .500 | 20.00 | 20.40 |
| 000000 | .469 | 18.75 | 19.12 |
| 00000 | .437 | 17.50 | 17.85 |
| 0000 | .406 | 16.25 | 16.57 |
| 000 | .375 | 15.00 | 15.30 |
| 00 | .344 | 13.75 | 14.02 |
| 0 | .312 | 12.50 | 12.75 |
| 1 | .281 | 11.25 | 11.47 |
| 2 | .266 | 10.62 | 10.84 |
| 3 | .250 | 10.00 | 10.20 |
| 4 | .234 | 9.37 | 9.56 |
| 5 | .219 | 8.75 | 8.92 |
| 6 | .203 | 8.12 | 8.29 |
| 7 | .187 | 7.50 | 7.65 |
| 8 | .172 | 6.87 | 7.01 |
| 9 | .156 | 6.25 | 6.37 |
| 10 | .141 | 5.62 | 5.74 |
| 11 | .125 | 5.00 | 5.10 |
| 12 | .109 | 4.37 | 4.46 |
| 13 | .094 | 3.75 | 3.82 |

## Table 5-45. Weights of Steel and Iron Plates and Sheets (U.S. Standard Gauge) (Cont'd)

| No. of Gauge | Thickness Inches (Approx.) | WEIGHT PER SQUARE FOOT | |
|---|---|---|---|
| | | Iron | Steel |
| 14 | .078 | 3.12 | 3.19 |
| 15 | .070 | 2.81 | 2.87 |
| 16 | .062 | 2.50 | 2.55 |
| 17 | .056 | 2.25 | 2.29 |
| 18 | .050 | 2.00 | 2.04 |
| 19 | .044 | 1.75 | 1.78 |
| 20 | .037 | 1.50 | 1.53 |
| 21 | .034 | 1.37 | 1.40 |
| 22 | .031 | 1.25 | 1.27 |
| 23 | .028 | 1.13 | 1.15 |
| 24 | .025 | 1.00 | 1.02 |
| 25 | .022 | 0.87 | 0.89 |
| 26 | .019 | 0.75 | 0.76 |
| 27 | .017 | 0.69 | 0.70 |
| 28 | .016 | 0.63 | 0.64 |
| 29 | .014 | 0.56 | 0.57 |
| 30 | .012 | 0.50 | 0.51 |
| 31 | .011 | 0.44 | 0.45 |
| 32 | .010 | 0.40 | 0.41 |
| 33 | .009 | 0.37 | 0.38 |
| 34 | .009 | 0.34 | 0.35 |
| 35 | .008 | 0.31 | 0.32 |
| 36 | .007 | 0.28 | 0.29 |
| 37 | .007 | 0.27 | 0.27 |
| 38 | .006 | 0.25 | 0.25 |

Courtesy Hoffman Specialty ITT

poses. It may be mechanically circulated with a blower system or it may be natural circulation through an open window.

*Vent Valve (Steam)*—A device for permitting air to be forced out of a heating unit or pipe and which closes against water and steam.

*Vent Valve (Water)*—A device permitting air to be pushed out of a pipe or heating unit but which closes against water.

*Warm-Air Heating System*—A warm-air heating plant consists of a heating unit (fuel-burning furnace) enclosed in a casing, from which the heated air is distributed to the various rooms of the building through ducts. If the motive head producing flow depends on the difference in weight between the heated air leaving the casing and the cooler air entering the bottom of the casing,

## Table 5-46. Areas of Circles

| Size | Area | Size | Area | Size | Area | Size | Area |
|------|------|------|------|------|------|------|------|
| $1/64$ | .00019 | $5^1/2$ | 23.758 | 25 | 490.87 | 64 | 3216.0 |
| $1/32$ | .00077 | 6 | 28.274 | 26 | 530.93 | 65 | 3318.3 |
| $1/16$ | .00307 | $6^1/2$ | 33.183 | 27 | 572.55 | 66 | 3421.2 |
| $3/32$ | .00690 | 7 | 38.484 | 28 | 615.75 | 67 | 3525.6 |
| $1/8$ | .01227 | $7^1/2$ | 44.178 | 29 | 660.52 | 68 | 3631.6 |
| $5/32$ | .01917 | 8 | 50.265 | 30 | 706.86 | 69 | 3739.2 |
| $3/16$ | .02761 | $8^1/2$ | 56.745 | 31 | 754.76 | 70 | 3848.4 |
| $7/32$ | .03758 | 9 | 63.617 | 32 | 804.24 | 71 | 3959.2 |
| $1/4$ | .04909 | $9^1/2$ | 70.882 | 33 | 855.30 | 72 | 4071.5 |
| $5/16$ | .07670 | 10 | 78.54 | 34 | 907.92 | 73 | 4185.3 |
| $3/8$ | .11045 | $10^1/2$ | 86.59 | 35 | 962.11 | 74 | 4300.8 |
| $7/16$ | .15033 | 11 | 95.03 | 36 | 1017.8 | 75 | 4417.8 |
| $1/2$ | .19635 | $11^1/2$ | 103.86 | 37 | 1075.2 | 76 | 4536.4 |
| $9/16$ | .24580 | 12 | 113.09 | 38 | 1134.1 | 77 | 4656.0 |
| $5/8$ | .30680 | $12^1/2$ | 122.71 | 39 | 1194.5 | 78 | 4778.3 |
| $11/16$ | .37122 | 13 | 132.73 | 40 | 1256.6 | 79 | 4901.6 |
| $3/4$ | .44179 | $13^1/2$ | 143.13 | 41 | 1320.2 | 80 | 5026.5 |
| $13/16$ | .51849 | 14 | 153.93 | 42 | 1385.4 | 81 | 5153.0 |
| $7/8$ | .60132 | $14^1/2$ | 165.13 | 43 | 1452.2 | 82 | 5281.0 |
| $15/16$ | .69029 | 15 | 176.71 | 44 | 1520.5 | 83 | 5410.6 |
| 1 | .7854 | $15^1/2$ | 188.69 | 45 | 1590.4 | 84 | 5541.7 |
| $1^1/8$ | .9940 | 16 | 201.06 | 46 | 1661.9 | 85 | 5674.5 |
| $1^1/4$ | 1.227 | $16^1/2$ | 216.82 | 47 | 1734.9 | 86 | 5808.8 |
| $1^3/8$ | 1.484 | 17 | 226.98 | 48 | 1809.5 | 87 | 5944.6 |
| $1^1/2$ | 1.767 | $17^1/2$ | 240.52 | 49 | 1885.7 | 88 | 6082.1 |
| $1^5/8$ | 2.073 | 18 | 254.46 | 50 | 1963.5 | 89 | 6221.1 |
| $1^3/4$ | 2.405 | $18^1/2$ | 268.80 | 51 | 2042.8 | 90 | 6361.7 |
| $1^7/8$ | 2.761 | 19 | 283.52 | 52 | 2123.7 | 91 | 6503.8 |
| 2 | 3.141 | $19^1/2$ | 298.64 | 53 | 2206.1 | 92 | 6647.6 |
| $2^1/4$ | 3.976 | 20 | 314.16 | 54 | 2290.2 | 93 | 6792.9 |
| $2^1/2$ | 4.908 | $20^1/2$ | 330.06 | 55 | 2375.8 | 94 | 6939.7 |
| $2^3/4$ | 5.939 | 21 | 346.36 | 56 | 2463.0 | 95 | 7088.2 |
| 3 | 7.068 | $21^1/2$ | 363.05 | 57 | 2551.7 | 96 | 7238.2 |
| $3^1/4$ | 8.295 | 22 | 380.13 | 58 | 2642.0 | 97 | 7389.8 |
| $3^1/2$ | 9.621 | $22^1/2$ | 397.60 | 59 | 2733.9 | 98 | 7542.9 |
| $3^3/4$ | 11.044 | 23 | 415.47 | 60 | 2827.4 | 99 | 7697.7 |
| 4 | 12.566 | $23^1/2$ | 433.73 | 61 | 2922.4 | 100 | 7854.0 |
| $4^1/2$ | 15.904 | 24 | 452.39 | 62 | 3019.0 | | |
| 5 | 19.635 | $24^1/2$ | 471.43 | 63 | 3117.2 | | |

To find the circumference of a circle when diameter is given, multiply the given diameter by 3.1416. To find the diameter of a circle when circumference is given, multiply the given circumference by .31831.

Courtesy Hoffman Specialty ITT

## Table 5-47. ASTM Schedule 40 (S) Pipe Dimensions

| DIAMETER | | | Nominal Thickness In. | CIRCUMFERENCE | | TRANSVERSE AREAS | | Length of Pipe per Sq. Ft. of External Surface Ft. | Length of Pipe Containing 1 Cu. Ft. | Nominal Weight Per Ft. in Lbs. | Nominal Number of Threads Per Inch of Screw |
|---|---|---|---|---|---|---|---|---|---|---|---|
| Nominal Internal In. | Actual External In. | Approx. Internal In. | | External In. | Internal In. | External Sq. In. | Internal Sq. In. | | | | |
| 1/8 | 0.405 | 0.27 | 0.068 | 1.27 | 0.85 | 0.13 | 0.06 | 9.44 | 2513.00 | 0.24 | 27 |
| 1/4 | 0.540 | 0.36 | 0.088 | 1.70 | 1.14 | 0.23 | 0.10 | 7.08 | 1383.30 | 0.42 | 18 |
| 3/8 | 0.675 | 0.49 | 0.091 | 2.12 | 1.55 | 0.36 | 0.19 | 5.66 | 751.20 | 0.57 | 18 |
| 1/2 | 0.840 | 0.62 | 0.109 | 2.63 | 1.95 | 0.55 | 0.30 | 4.55 | 472.40 | 0.85 | 14 |
| 3/4 | 1.050 | 0.82 | 0.113 | 3.30 | 2.59 | 0.87 | 0.53 | 3.64 | 270.00 | 1.13 | 14 |
| 1 | 1.315 | 1.05 | 0.134 | 4.13 | 3.29 | 1.36 | 0.86 | 2.90 | 166.90 | 1.68 | 11 1/2 |
| 1 1/4 | 1.660 | 1.38 | 0.140 | 5.22 | 4.34 | 2.16 | 1.50 | 2.30 | 96.25 | 2.27 | 11 1/2 |
| 1 1/2 | 1.900 | 1.61 | 0.145 | 5.97 | 5.06 | 2.84 | 2.04 | 2.01 | 70.66 | 2.72 | 11 1/2 |
| 2 | 2.375 | 2.07 | 0.154 | 7.46 | 6.49 | 4.43 | 3.36 | 1.61 | 42.91 | 3.65 | 11 1/2 |
| 2 1/2 | 2.875 | 2.47 | 0.204 | 9.03 | 7.75 | 6.49 | 4.78 | 1.33 | 30.10 | 5.79 | 8 |
| 3 | 3.500 | 3.07 | 0.217 | 11.0 | 9.63 | 9.62 | 7.39 | 1.09 | 19.50 | 7.57 | 8 |
| 3 1/2 | 4.000 | 3.55 | 0.226 | 12.57 | 11.15 | 12.57 | 9.89 | 0.96 | 14.57 | 9.11 | 8 |
| 4 | 4.500 | 4.03 | 0.237 | 14.14 | 12.65 | 15.90 | 12.73 | 0.85 | 11.31 | 10.79 | 8 |
| 5 | 5.563 | 5.05 | 0.259 | 17.48 | 15.85 | 24.31 | 19.99 | 0.69 | 7.20 | 14.62 | 8 |
| 6 | 6.625 | 6.07 | 0.280 | 20.81 | 19.05 | 34.47 | 28.89 | 0.58 | 4.98 | 18.97 | 8 |
| 8 | 8.625 | 8.07 | 0.276 | 27.10 | 25.35 | 58.43 | 51.15 | 0.44 | 2.82 | 24.69 | 8 |
| 8 | 8.625 | 7.98 | 0.322 | 27.10 | 25.07 | 58.43 | 50.02 | 0.44 | 2.88 | 28.55 | 8 |
| 9 | 9.625 | 8.94 | 0.344 | 30.24 | 28.08 | 72.76 | 62.72 | 0.40 | 2.29 | 33.91 | 8 |
| 10 | 10.750 | 10.19 | 0.278 | 33.77 | 32.01 | 90.76 | 81.55 | 0.36 | 1.76 | 31.20 | 8 |
| 10 | 10.750 | 10.14 | 0.306 | 33.77 | 31.86 | 90.76 | 80.75 | 0.36 | 1.78 | 34.24 | 8 |
| 10 | 10.750 | 10.02 | 0.366 | 33.77 | 31.47 | 90.76 | 78.82 | 0.36 | 1.82 | 40.48 | 8 |
| 12 | 12.750 | 12.09 | 0.328 | 40.06 | 37.98 | 127.68 | 114.80 | 0.30 | 1.25 | 43.77 | 8 |
| 12 | 12.750 | 12.00 | 0.375 | 40.06 | 37.70 | 127.68 | 113.10 | 0.30 | 1.27 | 49.56 | 8 |

Courtesy Hoffman Specialty ITT

it is termed a gravity system. A booster fan may, however, be used in conjunction with a gravity-designed system. If a fan is used to produce circulation and the system is designed especially for fan circulation, it is termed a fan furnace system or a central fan

## Table 5-48. Surface Areas and Volumes of Spheres

| Diameter | Surface Area | Volume | Diameter | Surface Area | Volume |
|---|---|---|---|---|---|
| ⅛ | .04908 | .00102 | 4¼ | 56.745 | 40.195 |
| ¼ | .19636 | .00818 | 4½ | 63.617 | 47.712 |
| ⅜ | .44180 | .02761 | 4¾ | 70.882 | 55.115 |
| ½ | .78540 | .06545 | 5 | 78.540 | 65.450 |
| ⅝ | 1.2272 | .12783 | 5¼ | 86.590 | 75.766 |
| ¾ | 1.7672 | .22090 | 5½ | 95.033 | 87.113 |
| ⅞ | 2.4053 | .35077 | 5¾ | 103.87 | 99.542 |
| 1 | 3.1416 | .52360 | 6 | 113.10 | 113.10 |
| 1⅛ | 3.9760 | .74550 | 6¼ | 122.72 | 127.83 |
| 1¼ | 4.9088 | 1.0227 | 6½ | 132.73 | 143.79 |
| 1⅜ | 5.9396 | 1.3611 | 6¾ | 143.14 | 161.03 |
| 1½ | 7.0684 | 1.7671 | 7 | 153.94 | 179.60 |
| 1⅝ | 8.2956 | 2.2467 | 7¼ | 165.13 | 199.53 |
| 1¾ | 9.6212 | 2.8062 | 7½ | 176.71 | 220.88 |
| 1⅞ | 11.045 | 3.4516 | 7¾ | 188.69 | 243.72 |
| 2 | 12.566 | 4.1887 | 8 | 201.06 | 268.08 |
| 2¼ | 15.904 | 5.9640 | 8¼ | 213.82 | 294.00 |
| 2½ | 19.635 | 8.1812 | 8½ | 226.98 | 321.55 |
| 2¾ | 23.758 | 10.889 | 8¾ | 240.53 | 350.77 |
| 3 | 28.274 | 14.137 | 9 | 254.47 | 381.70 |
| 3¼ | 33.183 | 17.974 | 9¼ | 268.80 | 414.40 |
| 3½ | 38.484 | 22.449 | 9½ | 283.53 | 448.92 |
| 3¾ | 44.179 | 27.612 | 9¾ | 298.65 | 485.30 |
| 4 | 50.266 | 35.511 | 10 | 314.16 | 523.60 |

### SPHERE FORMULA

This table can be used for feet, inches or any metric unit. For example, the volume of a 2″ diameter sphere is 4.1887 cu. in. and for a 2 ft. diameter, 4.1887 cu. ft. The figures apply to either the exterior or to the interior of a hollow sphere, provided the diameter is measured at the proper place. For example: the capacity of a spherical tank measuring 10 ft. on the inside is 523.60 cu. ft. A float ball having an outside diameter of 6 in. has a volume of 113.10 cu. in.

The area or the volume of a sphere of a diameter not given in the table may be figured from the following formula:

$$S = 4A \text{ and } V = 0.524D^3$$

in which:

$D$ = Diameter of the Sphere
$A$ = Area of a Circle of Diameter D
$S$ = Surface Area of Sphere
$V$ = Volume of Sphere

furnace system. A fan furnace system may include air washer, filters, and so on.

*Wet-Bulb Temperature*—The lowest temperature that a water-wetted body will attain when exposed to an air current.

*Wet Return (Steam)*—The part of a return main of a steam heating system that is completely filled with water of condensation.

*Wet Saturated Steam*—Saturated steam containing some water particles in suspension.

## Table 5-49. Decimal Equivalents of Parts of an Inch

| | | | |
|---|---|---|---|
| $1/64$ | .01563 | $33/64$ | .51563 |
| $1/32$ | .03125 | $17/32$ | .53125 |
| $3/64$ | .04688 | $35/64$ | .54688 |
| $1/16$ | .0625 | $9/16$ | .5625 |
| $5/64$ | .07813 | $37/64$ | .57813 |
| $3/32$ | .09375 | $19/32$ | .59375 |
| $7/64$ | .10938 | $39/64$ | .60938 |
| $1/8$ | .125 | $5/8$ | .625 |
| $9/64$ | .14063 | $41/64$ | .64063 |
| $5/32$ | .15625 | $21/32$ | .65625 |
| $11/64$ | .17188 | $43/64$ | .67188 |
| $3/16$ | .1875 | $11/16$ | .6875 |
| $13/64$ | .20313 | $45/64$ | .70313 |
| $7/32$ | .21875 | $23/32$ | .71875 |
| $15/64$ | .23438 | $47/64$ | .73438 |
| $1/4$ | .25 | $3/4$ | .75 |
| $17/64$ | .26563 | $49/64$ | .76563 |
| $9/32$ | .28125 | $25/32$ | .78125 |
| $19/64$ | .29688 | $51/64$ | .79688 |
| $5/16$ | .3125 | $13/16$ | .8125 |
| $21/64$ | .32813 | $53/64$ | .82813 |
| $11/32$ | .34375 | $27/32$ | .84375 |
| $23/64$ | .35938 | $55/64$ | .85938 |
| $3/8$ | .375 | $7/8$ | .875 |
| $25/64$ | .39063 | $57/64$ | .89063 |
| $13/32$ | .40625 | $29/32$ | .90625 |
| $27/64$ | .42188 | $59/64$ | .92188 |
| $7/16$ | .4375 | $15/16$ | .9375 |
| $29/64$ | .45313 | $61/64$ | .95313 |
| $15/32$ | .46875 | $31/32$ | .96875 |
| $31/64$ | .48438 | $63/64$ | .98438 |
| $1/2$ | .5 | 1 | 1.00000 |

## Table 5-50. Flue Sizes and Chimney Heights

| BOILER CAPACITY SQ. FT. RADIATION | | RECTANGULAR FLUE | | ROUND FLUE | | |
|---|---|---|---|---|---|---|
| Steam 240 Btu per Sq. Ft. | Hot Water 150 Btu per Sq. Ft. | Nominal Outside Dimensions of Fireclay Lining, In. | Effective Area Square Inches | Inside Diameter of Lining, Inches | Actual and Effective Area, Sq. In. | Height In Feet from Grate |
| 590 | 973 | 8½ × 13 | 70 | | | 35 |
| 690 | 1140 | | | 10 | 79 | |
| 900 | 1490 | 13 × 13 | 99 | | | |
| 900 | 1490 | 8½ × 18 | 100 | | | |
| 1100 | 1820 | | | 12 | 113 | 40 |
| 1700 | 2800 | 13 × 18 | 156 | | | |
| 1940 | 3200 | | | 15 | 177 | |
| 2130 | 3520 | 18 × 18 | 195 | | | |
| 2480 | 4090 | 20 × 20 | 234 | | | 45 |
| 3150 | 5200 | | | 18 | 254 | 50 |
| 4300 | 7100 | | | 20 | 314 | |
| 5000 | 8250 | 24 × 24 | 346 | | | 55 |

This table gives the minimum flue sizes allowable for steam or hot water systems. In connection with this table, the following points should be noted:
1. The ratings given are based on smooth, lined, straight flues. If there is more than one offset or if any single offset is flatter than 60° with the horizontal, the rating should be reduced accordingly.
2. Flue heights and sizes are based upon an altitude not over 500 feet above sea level. For each one thousand feet above sea level, the height of the flue for a given appliance should be increased 10% above the figures in the table and capacities in the table should be reduced 5%.
3. The effective area of an unlined round flue shall be based upon a diameter two inches less than the actual diameter. The effective area of an unlined rectangular flue shall be computed from the formula:

$$\text{Effective Area in Square Inches} = \frac{\pi(B - 2)^2}{4} + (A - B)(B - 2)$$

in which (A) and (B) are, respectively, the long and short inside dimensions in inches.
4. When two appliances are to be connected to one flue, the rated capacities of the flue of various sizes shall be taken as 60% of those given in the table. When three appliances are to be connected, the capacities shall be taken as 40% of those given in the table. When four appliances are to be connected, the capacities shall be taken as 35% of those given in the table.
5. The flue heights and szes are for ordinary updraft boilers. For special types of boilers, greater heights and area may be required.

**Courtesy Hoffman Specialty ITT**

213

## Table 5-51. Chimney Data

POOR DRAFT

GOOD DRAFT

AN IDEAL TOP
FOR CHIMNEY
(A)

(B)

GOOD DRAFT

POOR DRAFT

CHIMNEY BUILT UP ABOVE
ADJOINING BUILDING TO
OBTAIN GOOD DRAFT
(C)

(D)

SHAPES OF FLUES

(E)

(F)

(G)

(H)

IDEAL

GOOD

FAIR

POOR

EXAMINING FLUE
WITH MIRROR
(I)

SMOKE PIPE CUTTING
OFF THE DRAFT
(J)

CORRECT
CONSTRUCTION
(K)

**Courtesy Hoffman Specialty ITT**

# COMMON HEATING TROUBLES IN STEAM HEATING SYSTEMS

The following are a few of the more common difficulties experienced with steam or hot-water heating systems, and the probable causes.

## Boiler Troubles

*The Boiler Fails to Deliver Enough Heat*—The cause of this condition may be: (a) poor draft, (b) poor fuel, (c) improper attention or firing, (d) boiler too small, (e) improper piping, (f) improper arrangement of sections, (g) heating surfaces covered with soot, and (h) improper firing rate of oil or gas burner.

*The Water Line Is Unsteady*—The cause of this condition may be: (a) grease and dirt in boiler, (b) a water column connected to a very active section and therefore not showing actual water level in boiler, and (c) boiler operating at excessive output.

*Water Disappears from Gauge Glass*—This may be caused by: (a) priming due to grease and dirt in boiler; (b) too great a pressure difference between supply and return piping, causing water to back into return; (c) valve closed in return line; (d) connection of bottom of water column into a very active section or thin water-way; (e) improper connections between boilers in battery, permitting boiler with excess pressure to push water into boiler with lower pressure; and (f) too high a firing rate.

*Water Is Carried into Steam Main*—This may be caused by: (a) grease and dirt in boiler, (b) type of boiler not adapted to job, (c) outlet connections too small, (d) using boiler beyond rated capacity, (e) water level carried higher than required, and (f) firing rate too high.

*Boiler Is Slow to Respond*—This may be due to : (a) poor draft, (b) inferior fuel, (c) improper attention, (d) accumulation of clinkers on grate, (e) boiler too small for the load, and (f) improper firing rate.

*Boiler Flues Collect Soot Quickly*—This may be due to: (a) poor draft; (b) smoky combustion; (c) too low a rate of combustion; (d) excess of air in firebox, causing chilling of gases; and (f) improper firing rate.

*Boiler Smokes through Fire Door*—This may be due to: (a) defective draft in chimney or incorrect setting of dampers, (b) air leaks into boiler or breeching, (c) gas outlet from firebox plugged with fuel, (d) dirty or clogged flues, and (e) improper reduction in breeching size.

*Pressure Builds up Very Quickly (as indicated by gauge) But Steam Does Not Circulate*—This is due to grease and dirt in boiler.

215

## Piping Troubles

Water hammer is one of the chief causes of noise in steam heating systems and also is the cause of much damage to traps, vents, and the like. It is a wave transmitted through a pipe filled, or partially filled, with water. It may have its origin in waves set up by steam passing at a high velocity over condensate collected in the piping. Fig. 5-1 illustrates how such a wave may be formed. The rapid passage of steam over the surface of the water causes a wave to form, as at B. The rapid condensation of the steam in pockets A brings the two slugs of water together with a considerable force, which may be telegraphed through the piping. If this wave is intercepted by, say, the float of a vent valve, not only will noise result but damage to the float in the valve is almost certain.

Courtesy Hoffman Specialty ITT

Fig. 5-1. Water hammer.

*Water Hammer in Hartford Connection*—In the Hartford connection a close nipple should be used between the end of the return and the header drip or equalizing pipe. If the nipple used at this point is too long and the water line of boilers becomes low, hammer will occur. The remedy is to offset the return piping considerably below the water line of the boiler, so that a close nipple can be used in entering header drip and maintain proper boiler water line. Top of close nipple should be 2″ below water line.

*Water Hammer in Mains*—

1. Water pocket formed by sagging of the main. An easy and reliable method to check this is to stretch a chalk line between fittings and note relation of the pipe to the line.
2. Improper pitch of mains. Check with level.
3. Water hammer is also caused by too great a pressure drop, due to insufficient pipe sizes, unreamed piping, or other restrictions in the line.

4. Insufficient water-line difference between the low point of the horizontal main and boiler water line. In one-pipe gravity systems, this distance should normally be 18' or more and in vapor systems, the ends of dry return mains should be 24" or more above water line, depending on size of installation and pressure drop.
5. Improper location of air valves for venting steam main (one-pipe gravity or vacuum systems).
6. Excessive quantities of water in main due to priming boiler or improper header construction. All boiler tappings should be used and connected full size to boiler header.

*Steam Does Not Circulate to Ends of Mains—*

1. Dirty boilers.
2. Insufficient water-line difference.
3. Improper venting of mains.

## Radiator Troubles

*Pounding (One-Pipe System)—*

1. Radiator supply valve too small or partially closed.
2. Radiator pitched away from supply valve.
3. Vent port of air valve too large, allowing steam to enter radiator too rapidly.

*Radiator Does Not Heat—*

1. Improper venting of air.
2. Branch supply to radiator too small.
3. Vent port of air valve clogged with dirt (one-pipe systems).
4. Drainage tongue of air valve damaged or removed.
5. Branch supply improperly pitched, causing water pocket.
6. Steam pressure higher than maximum working pressure of vent valve. This is especially likely to happen where steam is supplied through a reducing valve from high-pressure supply.
7. Return branch improperly pitched, causing water pocket to form, trapping air (vapor system).
8. In one-pipe vacuum system, if gas or oil-fired, some radiators may not heat if on previous firing they were not completely

217

heated. Changing to an open (nonvacuum) system by using open vents on radiators and mains would remedy this.

*Radiator Cools Quickly (One-Pipe Vacuum System)*—

1. Air leakage into system either through leaky joints or through stuffing box of radiator supply valve if ordinary valve is used.
2. Dirt in vacuum valve preventing formation of vacuum.
3. On gas or oil-fired system, radiator cools quickly due to rapid formation of vacuum, and it is better to change to open (nonvacuum) system by changing radiator and main vents to nonvacuum (open) type.

## Reaming and Pitching of Piping

Many carefully laid out systems are found to have radiators that cannot be completely heated under test. Improper reaming of pipes can be given as the cause of many such conditions. In mains using unreamed pipe, water pockets are formed and frequently cause trouble.

In many cases troubles caused by burrs can be overcome by repitching the pipe line. If a line having ¼" pitch in 10' is increased to ½" in 10', the capacity is increased 20% in ¾" and 1" lines and 10 to 12% in 1¼" to 2". See Comparative Capacity of Steam Lines at Various Pitches in Section 7. The selection of pipe with smooth interior surface is likewise important in its effect on carrying capacity.

## Boiler Cleaning

After a new installation has been in service for a week or so, grease, oil, scale, core sand, and the like will accumulate in the boiler. It should then be thoroughly cleaned with some boiler-cleaning compound. There are various cleaning compounds on the market, and it is important to follow the recommendations of the boiler manufacturers in the selection of the proper one and to follow closely the prescribed method of using it.

Air valves should not be installed in a new system prior to the cleaning of the boiler. If temporary heat is required, the air may be vented with pet cocks or with old vent valves.

In systems using vacuum or condensation pumps, the return to pump should be closed off and all condensation passed to drain until the boiler has been cleaned.

CHAPTER 6

# Hot-Water
# Heating Systems

Water has been used as a medium for transmitting heat for many years. The Btu (British thermal unit) is the unit of measurement of heat. Radiation and convectors are designed to furnish a given number of Btu's at specified water temperatures. Hydronics describes hot-water heating systems or combination hot-water and chilled-water heating and cooling systems.

Heat is transmitted in three ways: by conduction, convection, and radiation.

*Conduction* is the process of transferring heat from one molecule to another. If one end of an iron bar is heated in a flame, the heat will travel to the other end by conduction.

*Convection* is the process of transmitting heat by means of matter. The matter can be either a liquid or a gas (air). Fig. 6-1 illustrates how air is heated by convection. As the radiator is heated, the air surrounding it is also heated. As the air is heated it

221

Approximately 40% of the heat effect of cast iron radiation is in the form of infrared waves. convection, the transfer of heat from the radiator to the air flowing around it, accounts for approximately 60%.

Approximately 80% of the heat effect of cabinet convectors is due to convection, only 20% is due to infrared waves.

∿∿→   INFRARED waves

∿∿→   CONVECTION

**Fig. 6-1. The heating effect of cabinet convectors and cast-iron radiators.**

also becomes lighter and rises, and as it rises it gives off heat. Circulation is established when the heated air cools and drops, flows toward the heating surface, and again rises.

*Radiation* is the process of transferring heat by means of heat waves. Heat waves are similar to radio waves, the difference being the frequency or length. The earth is heated by the sun, the rays or waves travel through the vacuum of space to warm the earth. The terms *latent* heat and *sensible* heat are often used when working with hydronic systems. Fig. 6-2 illustrates the meaning of these terms.

## HOT-WATER HEATING BOILERS

Hot-water heating boilers are designed to operate at a pressure not to exceed 30 psig. Cast iron is the most widely used material for small hot-water boilers; large boilers are usually made of steel. The sections of a cast-iron sectional boiler can be moved up and down stairwells, through doorways, and into and out of areas that would be impossible to negotiate with a steel boiler. Cast-iron boilers can be fired with any of the conventional fuels. Metal cabinets or jackets can be installed to enclose the boiler and its insulation, cutting down on heat loss from the boiler and presenting a pleasing

ICE TEMPERATURE
32° F

Ice is water in a solid state. When latent heat is added, the ice is converted to a liquid state. As this stage is reached, unless further heat is added, the water remains at 32° F.

WATER TEMPERATURE 32°

WATER ABSORBS OR ASSIMILATES SENSIBLE HEAT

As the water absorbs heat from the burner, the water temperature rises. This absorbed heat is called sensible heat.

WATER CHANGES FROM A LIQUID TO A GASEOUS STATE (STEAM) AT 212° F AT SEA LEVEL

The water continues to absorb sensible heat until it reaches a temperature of 212°F (at sea level). At 212°F the water absorbs latent heat and is changed from a liquid to a gaseous state called steam. Under these conditions the steam temperature remains at 212°F.

**Fig. 6-2. The effects of latent and sensible heat.**

finished appearance. Hot-water heating systems can vary from a simple one-pipe gravity system to a highly sophisticated zoned reverse-return system. Controls can be installed that will raise or lower the boiler water temperatures as the outside temperature fluctuates.

Water in a gravity heating system circulates because of the difference in density, and therefore a difference in weight, of hot and cold water. One cubic foot of water at a temperature of 200°F weighs 60.13 lbs. If a container holding one cubic foot of water at 200°F is cooled to 52°F it will be found that the water has decreased in volume, or become denser. More water will have to be added to the container to bring the level up to the one cubic foot level and the net weight of the water will then be 62.42 lbs. As water is heated it becomes lighter or less dense; as it cools it becomes heavier or more dense. Below 52°F, no further shrinkage takes place; at 32°F, just before freezing, water still weighs 62.42 lbs. per cubic foot.

Fig. 6-3 shows the circulation by gravity of a hot-water system. The radiation cools the supply of hot water from the boiler by giving off heat. As the water from the radiator enters the return main, being more dense and therefore heavier than the hot water in the boiler, it pushes the hot water up and out of the boiler and establishes circulation. Gravity hot-water systems are now out-moded and rarely installed.

Fig. 6-4 shows a typical one-pipe forced-circulation hot-water heating system. Note that the supply main is reduced in size between the tees (B), to force part of the supply to go up and through the first radiator. At point (C), between the radiators, the supply main is again full size. On a cold day with the radiators giving off heat, there could be a marked difference in the tempera-ture of the supply-main water between points (A) and (C). Radia-tor No. 2 has more heating surface because of the two windows but may have to be further increased in size because of the differ-ential in supply water temperature. At point (D) the supply main is again reduced in size to force circulation through radiator No. 2. At point (E) the return main begins and is again full size. Fig. 6-5 shows this installation using reducing tees at the supply- and return-main connections. Balancing cocks may be needed on the returns from the radiators shown in Fig. 6-5 to ensure adequate

VENT   VENT AND OVERFLOW

EXPANSION TANK

WATER-LEVEL GAUGE

OVERFLOW

CIRCULATING PIPE

AIR VENT

RETURN MAIN

FLOW MAIN

WATER SUPPLY

BOILER

DRAIN

Courtesy Dunham-Bush Inc.

Fig. 6-3. A typical gravity pipe system. The flow of water takes place because of the difference in weight between the hot water in the supply main and the cooler water in the return main.

**Fig. 6-4. A typical one-pipe, forced-circulation, hot-water heating system.**

flow through each radiator. This system can also be designed using flow fittings (tees) made with orifices and diverters, instead of the cast-iron fittings (tees) shown. Diverter fittings take advantage of the stratification of water, as shown in Fig. 6-6.

226

## DIRECT-RETURN SYSTEMS

In a two-pipe direct-return system, shown in Fig. 6-7, the supply main begins at the first radiator, the one closest to the boiler or supply source, and ends at the last radiator. The return main starts

Fig. 6-5. A typical two-pipe, forced-circulation, hot-water heating system.

HOTTEST WATER AT TOP

COOLEST WATER AT BOTTOM

**Fig. 6-6. Water stratification in a one-pipe, hot-water heating system.**

at the last radiator and returns to the boiler or supply source, connecting to the return of each radiator, with the result that each radiator has a different length of supply and return piping. The first radiator off the supply main has the shortest return path back to the boiler or source, while the last radiator on the supply piping has the longest path back. This creates balancing problems, and the system must be provided with balancing cocks or valves on the return side of each radiator. The balancing cocks may need to be adjusted very accurately to ensure flow, and heat, to the farthest radiator in the system.

## REVERSE-RETURN SYSTEMS

A two-pipe reverse-return system is shown in Fig. 6-8. The reverse-return system has many advantages over the direct return. In the reverse-return system the first radiator supplied from the supply main has the shortest supply main but has the *longest* return main. The supply main to each radiator in the system is progressively longer, but as the supply main is longer, the return main is progressively shorter. The result is that the actual developed length of supply and return main to each radiator is basically the same, or balanced. The initial cost of installation of a reverse-return system may be somewhat higher than for a direct-return system, due to the extra labor and material involved, but the advantages of a reverse-return system are well worth the extra cost.

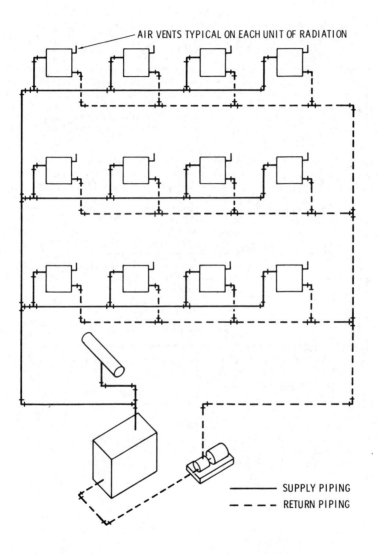

**Fig. 6-7. Two-pipe, direct-return system using base-mounted circulating pump.**

AIR VENT TYPICAL ON EACH UNIT OF RADIATION

ZONE 3

ZONE 2

ZONE 1

SUPPLY PIPING ZONE 1
SUPPLY PIPING ZONE 2
SUPPLY PIPING ZONE 3

CIRCULATING PUMP FOR ZONE 1
CIRCULATING PUMP FOR ZONE 2
CIRCULATING PUMP FOR ZONE 3

——— SUPPLY
— — — RETURN

**Fig. 6-8. Reverse-return system with booster pumps for each zone.**

# AIR CONTROL IN HOT-WATER HEATING SYSTEMS

Air is necessary in the proper operation of a hot-water heating system because of the fact that water is virtually incompressible. In a closed hot-water heating system the water in the boiler or heat exchanger expands as it is heated. In order to provide space for the expansion of the water, an expansion, or compression, tank must be provided. The location of the expansion tank is not critical; it is common practice to install expansion tanks adjacent to and above the boiler or heat exchanger. A expansion tank will normally contain from ⅓ to ½ water, with air space above the water.

For proper operation of a hot-water heating system, the air cushion in the expansion tank should be the only air in a closed hydronic system. Because air can be absorbed into water, some method must be used to continuously separate the free air from the water in the system and conduct it to the expansion tank, and at the same time restrict the flow of water from the expansion tank (gravity circulation) without restricting the passage of free air into the tank. A tank fitting, shown in Fig. 6-9, is designed for this purpose. Another fitting is used on expansion tanks of 100 gals. and larger (shown in Fig. 6-10).

**Courtesy Bell & Gossett ITT**

**Fig. 6-9. Tank fitting that separates air from water.**

An in-line air separator is shown in Fig. 6-11. It is used to separate the air from the water in heating and/or cooling systems. The use of this type of fitting eliminates the need for automatic air vents and prevents waterlogged expansion tanks. Waterlogging results when a compression or expansion tank becomes filled with water, leaving no air space for the expansion of water in the system.

Another type of in-line fitting is shown in Fig. 6-12. This control

231

Fig. 6-10. Tank fitting designed for use in expansion tanks of 100 gals. and larger.

Fig. 6-11. Air separator.

232

is very effective and can be used when the boiler or converter will not accommodate a dip tube for the purpose of air separation.

Side or end outlet boilers are generally tapped for forced-circulation pipe sizes. A boiler fitting control can be used on this type of boiler to aid in removing free air from the boiler water. This control is shown in Fig. 6-13.

**Courtesy Bell & Gossett ITT**

**Fig. 6-12. In-line fitting.**

**Courtesy Bell & Gossett ITT**

**Fig. 6-13. Boiler fitting control.**

The boiler fitting shown in Fig. 6-14 is designed for use on top outlet boilers.

In summation, the elimination of air from the boiler or converter and from the piping system is vitally important in order for the system to function as designed and remain troublefree.

233

**Fig. 6-14. Boiler fitting for top-outlet boilers.**

## AIR VENTS

Free air can cause noise in the piping system and can also interfere with water circulation. On the start-up of a heating system, either because it is new and dry or because it has been drained, the air that will be forced to the high points in the system must be removed. Manually operated air vents should be installed at high points in the piping, if these points are readily accessible; otherwise, automatic air vents should be installed at these points. A drain line ( ¼" O.D. copper tubing) should be installed from the automatic vents to a suitable receptor. Some of the common manual and automatic air vents are shown in Fig. 6-15. When the manually operated air vent is opened, it should be left open until a steady stream of water is obtained. It may be necessary to check all the air vents several times to purge all the air from the system. When a hot-water heating system is placed in operation, hot water is circulating through the system, and the system has been properly balanced; if one or more units of radiation do not heat it is generally because air is trapped in the unit. Venting the air will relieve this problem.

## HOT-WATER BOILER WATER FEEDERS

Water feeders of the type shown in Fig. 6-16 are used to supply make-up water to hot water heating boilers. The operating range of this type valve is adjustable to suit varying building heights. The valve shown is designed for fast feeding of a boiler and piping system.

(A) Manually operated air vent.

(B) Manual and/or automatic air vent.

(C) Automatic air vent.

Courtesy Bell & Gossett ITT

**Fig. 6-15. Some common manual and automatic air vents.**

**Fig. 6-16. Feed-pressure water regulator.**

Courtesy Watts Regulator Co.

235

# EXPANSION TANKS

When a heating system has been filled with water and the expansion (compression) tank is in normal operating condition, the tank will be from ⅓ to ½ full of water; the balance of the tank will contain air. The air in an expansion tank serves a a cushion for thermal expansion of water in the system. The tank must be absolutely airtight; if there is a leak in the tank, the air will escape and be replaced with water. An expansion tank filled with water is said to be waterlogged. An expansion tank should be carefully checked at the time it is installed and periodically thereafter, with particular attention given to the gauge cocks and the packing nuts securing the gauge glass.

When a boiler and heating system is placed in operation, with cold water in the system, the water volume will expand when the water is heated. From 40° to 200° water will expand to 1.04 its original volume. Assuming that the expansion tank has been sized correctly, there will be sufficient air space in the expansion tank to accommodate this increased volume. As the water in the boiler is heated it expands, and as it expands it builds up pressure. The air cushion in the tank will accommodate the increased pressure, up to 30 lbs., at which point the relief valve on the boiler will open and relieve the built up pressure. A combination temperature, pressure, and altitude gauge should be installed on the boiler in order to check operating pressures and temperature. When filling a system, or in normal operation, pressure should not be excessive; 15 to 18 lbs. should be satisfactory and will allow for water volume expansion and pressure build up. When fast filling a system through a manual fill valve, excess pressure can be relieved by opening the relief valve momentarily. In a closed system if the expansion tank becomes waterlogged, with no space for the expanded water volume, the pressure will build up very quickly as the water is heated, and the relief valve will open and relieve the excess pressure. When this condition is encountered the reason for the expansion tank becoming waterlogged must be found and corrected. A typical expansion tank is shown in Fig. 6-17.

The expansion (compression) tank shown in Fig. 6-17(A) is typical of expansion tanks used in hydronic systems. The bottom side of the tank is shown; in use the openings shown would be

Courtesy Bell & Gossett ITT

(A) A typical compression (expansion tank used with hot water heating systems).

Courtesy Watts Regulator Co.

(B) A boiler water safety relief valve.

Fig. 6-17. A typical compression (expansion) tank used with hot-water heating systems.

turned down. An expansion tank should also have openings on one end of the tank for gauge cocks and a gauge glass. To help prevent air leakage leading to waterlogging of the tank, the tank should have no openings on the top side.

## HOT-WATER BOILER RELIEF VALVES

Boiler water safety relief valves are installed on boilers for protection against excessive water pressure caused by the thermal expansion of water. The type shown in Fig. 6-17 (B) has an emergency Btu steam-discharge capacity if runaway firing conditions should occur.

## HOT-WATER BOILER
## OPERATING CONTROLS

Two aquastats should be installed in a hot-water boiler, one to serve as the operating control, the other to serve as a high-limit control. The operating control will maintain a set minimum water temperature and the high-limit control will inactivate the heat source if the boiler water temperature exceeds the set point of the high-limit control. Both controls are actually thermally operated electrical switches. The operating control is a normally open switch that makes contact or closes a circuit on temperature fall. When the water temperature in the boiler is below the set point of the aquastat, the switch in the control will be in closed position, calling for heat. The high-limit control is a normally closed switch that opens and breaks contact, breaking the electrical circuit, when the boiler water temperature rises above the set point of this control. Both aquastats are wired in series with the gas valve or other heat source. Other safety controls may also be installed on a boiler, in which case they would also be wired in series with the operating control and the high-limit control. Opening a switch, or breaking the electrical circuit, of any single unit in a series wired circuit will interrupt the flow of current in that circuit and inactivate the heat source. Operating and high-limit aquastats are shown in Fig. 6-18.

**Fig. 6-18. The high-limit aquastat is normally set 10° above the operating aquastat.**

## CIRCULATING PUMPS FOR HOT-WATER HEATING SYSTEMS

Two types of pumps are in general use for circulating the water in hydronic heating and cooling systems. Booster pumps of the type shown in Fig. 6-19 are generally used on smaller systems for individual zone circulation, while the larger, base-mounted pumps, similar to the type shown in Fig. 6-20 are used on larger piping systems.

239

Courtesy Bell & Gossett ITT

Fig. 6-19. Booster pump.

Courtesy Bell & Gossett ITT

Fig. 6-20. Base-mounted circulating pump.

240

It is considered good piping practice to install valves on the inlet and outlet sides of the circulating pumps in hot-water heating systems. Gate valves can be used with booster pumps to isolate the pumps for repair or replacement, and to hold the water in the system and thus avoid the air problems that would arise if a zone or a system required refilling.

The discharge side of a centrifugal pump used in a heating and/or cooling system should be provided with a shut-off valve, a balancing valve to regulate the flow, and a check valve. The valve shown in Fig. 6-21 is designed to perform all three functions.

It is a check valve and a shut-off valve, and it is made with a calibrated stem for returning the valve to the "set" position after the valve is used as a shut-off. When the valve is in the "set" position, it serves as a balancing valve.

Courtesy Bell & Gossett ITT

**Fig. 6-21. Triple-duty valve.**

The piping shown in Fig. 6-22 is typical of most baseboard, fin-tube convector, and cabinet heater piping. Some systems include an automatic temperature control valve. This, if used, would be located between the supply valve and the radiation. The square head balancing cock is used to pinch down the flow through the radiation in order to balance the system.

**Fig. 6-22. Typical baseboard radiation connections.**

## RADIANT-PANEL HEATING

Radiant-panel heating is a method of heating a room by raising the temperature of one or more of its interior surfaces (floor, walls, or ceiling), as in Fig. 6-23, instead of heating the air, as in Fig. 6-24. Conventional heating systems heat the air within the room by convection, giving off very little radiant heat. The name *radiant-panel heating* is derived from the fact that tubes or pipes are placed in the floor, walls, or ceiling to form a heating "panel."

There are various methods used to achieve radiant heating. One of the most common, and the one that relates to the plumbing trade, is where hot-water pipes or tubes are embedded in the structure of the building. Other methods of radiant heating are:

1. By circulating warm air through shallow ducts beneath the floor or in spaces formed in the walls.
2. By the use of electrically heated wires, metal plates, or glazed fire-clay panels.
3. By the use of electrically heated tapestry formed into portable screens or hung on the walls of a room.

242

74° 73° 72° 69° 67°

CONVENTIONAL HEATING

**Fig. 6-23. Temperature levels in a room heated by a conventional radiator (or warm-air register). Notice the wide variation in temperature. Air currents often create drafts, spread dust, and cause hot air to collect at the ceiling and cool air at the floor.**

4. By attaching fabricated metal plate to walls and ceiling, through which hot water or steam is circulated.

The installation procedure generally accepted for radiant-panel heating is to embed specially constructed pipe coils or lengths of tubing in the floor, walls, or ceiling. These coils generally consist of small-bore wrought-iron, steel, brass, or copper pipe, usually with an inside diameter of $\frac{3}{8}''$ to $1''$.

## Insulation Requirements

Extra insulation is unnecessary, at least in terms of comfort, when radiant-panel heat is employed. In fact, a poorly insulated house that is often uncomfortable because of the prevailing low temperature of the walls will be much more comfortable when radiant heat is installed. This is because the inner surfaces of the outside walls will be at a higher temperature than before. Two

243

RADIANT-PANEL HEATING

**Fig. 6-24. Radiant-panel heating warms the interior surfaces (ceilings, walls, floor) of a room instead of the air. The result is the elimination of cold, drafty floors and a more even temperature from floor to ceiling.**

main factors should be considered, however, before a decision is made to convert to radiant-panel heating.

1. If a house has only fair insulation, comfort may dictate that a radiant system be installed in the ceiling only; floor systems will have an excess output under these conditions and will be uncomfortably warm.

2. Consider that insulation will always pay for itself in time by the savings in fuel.

### Air Venting

Air venting is an essential feature in the control of any panel-heating system. Any small amount of air collecting in either the circuit pipes or pipe coils will cause continued trouble and invariably result in a shortage of heat. An arrangement similar to the one in Fig. 6-25 will permit uniform venting. Because of the continuous slope to the coil connections, it may prove sufficient to install

**Fig. 6-25. A radiant-panel heating system showing draining and venting.**

automatic vents at the top of the return riser only, omitting such vents on the supply riser.

The fill and drain lines in Fig. 6-25 are connected at the boiler. The difference in elevation $H$ should be at least 3″, or sufficient to provide for a continuous slope to the coil connection. This slope should not be excessive, however, in order to fully utilize the gravitational forces acting on the system when heated.

## System Leakage Test

No troubles should be experienced with leaks in a radiant-heating system if the proper procedure is followed and reasonable care taken in installing. After the installation is completed, a hydraulic pressure test, as recommended by the tube or pipe manufacturer, is applied to the system.

The material immediately surrounding the coil protects it from

245

corrosion. An additional safety measure is to connect the coils to the water-supply source during the installation period. Thus, it will immediately be noticed if the piping becomes damaged.

## Panel Temperatures

The temperature conditions that are satisfactory for comfort by the occupant determine the highest allowable panel surface temperature. For rooms occupied continuously, the floor temperatures should never be above 85°F. For ceilings, the maximum allowable temperature depends on the height. Generally, 100°F is considered maximum for 7' ceilings, with 10°F added for each additional foot. Thus, the maximum allowable temperature of a 9' ceiling is 120°F. With the coils embedded in plaster, the maximum temperature is considered to be 130°F.

The temperature drop of the water in the panels should be limited to maintain the efficiency of the heating system. For practical purposes, the temperature difference between the inlet and outlet water should be between 15° and 30°F for floor panels, and between 20° and 35°F for ceiling panels.

## Advantages

The advantages of radiant-panel heating are numerous. Among the most common are:

1. More floor space. Radiant-panel heating eliminates radiators and grilles, as well as the need for window recesses to accommodate radiators. Elimination of radiators actually provides more floor space.
2. Interior of rooms can be better decorated. With radiant-panel heating, no restrictions exist as to furniture arrangement and wall decorations.
3. Lower velocities of air currents and therefore less streaking and dust on walls and ceilings.
4. Warm floors in homes without basements.
5. Simplifies troublesome architectural and engineering design problems.
6. Lower operating costs. A well-designed radiant-panel heat-

ing system should result in the lowest overall operating and general maintenance expense for the heating plant.

## Installation

Radiant-heating coils can be installed in the floor, ceiling, or walls. Ceiling installation provides a greater percentage of radiant-heat transmission than either floors or walls. For example, only about 50% of the heat emitted from floor panels is in the form of radiant rays, while about 65% of the heat emitted by ceiling panels is radiant. The percentage for wall panels is even less than for floor panels.

The shape and size of the coils used in hot-water radiant-heating systems depends on such factors as pressure drop, the shape of the space to be heated, and the material of which the pipes are made. For reasons of economy, it is important to lay out the lines in long, straight runs. In addition, it is necessary to select panel arrangements that result in the best comfort conditions for the rooms to be heated. In most cases, it is advisable to locate the panel close to outside walls and entrances. In rooms having the greatest heat loss, it may be necessary to spread the heating coil over the entire ceiling or floor area. In other rooms, possibly one ½ to ¾ of the ceiling or floor area will need to be covered by the heating coil.

Copper tubing is the material most often used in the installation of hot-water radiant heating. A minimum of solder joints should be used, so short sections of tubing should be avoided if possible. Instead, continuous coils should be used, as shown in Fig. 6-26.

The shape of the radiant-heating coil will usually be determined by the size and shape of the area to be heated. Fig. 6-27 shows a conventional sinusoidal coil shape. Here, the area immediately inside the door is not covered by the coil since the occupants will not be sitting or standing in this part of the room.

The tube or pipe spacing should be restricted to 12″ or less for all radiant-heating panels. Spacings greater than this will result in noticeable temperature variations between various points on the surface of the panel.

Fig. 6-28 illustrates a grid-type coil for use where a low-pressure drop is desired or where a constant temperature of the entire panel surface is required. Irregularly shaped rooms often require odd-

Fig. 6-26. Copper tubing for a radiant floor panel.

Fig. 6-27. A conventional sinusoidal radiant-heating coil.

shaped coils, such as the one in Fig. 6-29. In large rooms where several coils must be used, a combination grid and continuous coil system (Fig. 6-30) is often used. In this type, common supply and return headers are connected to the mains.

Hot-water radiant panels can be installed in nearly any home— with or without a basement. Conventional hot-water boilers are used. These units are available in compact types that fit into closets or small utility rooms and are fired by gas or oil. Additions can be installed at any time, provided the limitations of the boiler unit and circulating pump are not exceeded.

The maximum length of coils is determined by many factors, but in most installations should be limited to 200' for ⅜" tubing in

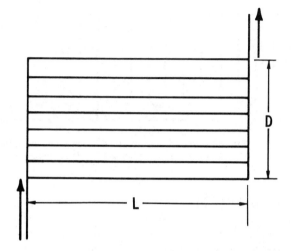

**Fig. 6-28. A grid-type heating coil used where a low-pressure drop is desired or a constant panel-surface temperature is required.**

ceiling and wall panels. Length of floor coils should be limited to 180', 280', and 450' for ½", ¾", and 1" tubing, respectively.

Radiant panels should never be used with steam instead of hot water, as too many complications arise. Also, domestic hot water should not be taken from the system for use in bathrooms or kitchens. The addition of fresh water should be kept to a minimum to avoid accumulation of impurities in the water.

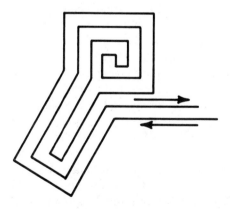

**Fig. 6-29. A coil for an irregularly shaped room.**

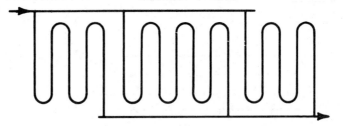

**Fig. 6-30.** A combination grid and continuous coil for use in large installations where the heat requirements make it necessary to install several coils.

Provision should be made for draining the system if the need should arise. Care therefore must be taken in the design and installation to ensure that the system can be completely drained, with no water pockets existing that will hold water and cause damage in the event of a freeze. The system should be installed with a vent at the high point and a drain at the low point, and without traps so that it may be easily drained.

Cold water should never be circulated in the system during hot weather in an attempt to provide cooling. Sweating of the panels will almost certainly take place, resulting in damage or generally unsatisfactory conditions.

**Ceiling Panels.** The most satisfactory location of radiant-heating panels is in the ceiling. These are usually fastened to the underside of the lath and then covered and embedded in plaster as the room is finished. If properly installed, this type of installation will not cause the plaster to crack, provided the plaster is allowed to cure and dry thoroughly before hot water is introduced into the coil. Insulation should be provided above the ceiling to prevent excessive heat transfer into the attic or rooms above.

Wallpaper or paint can be applied to the plaster in which the heating coils are embedded. The wallpaper or paint should be allowed to dry thoroughly before heating is attempted.

A typical piping diagram for a ceiling panel is shown in Fig. 6-31. In this arrangement, the hot water enters the panel near the outside walls and leaves from the inside. The control valves may be located in either the supply or return line to suit the convenience of the particular installation. The surface of the lath should be as

250

**Fig. 6-31. A typical ceiling panel.**

level as possible. The coils may be preformed and lifted into place, or may be formed directly on the ceiling.

Forming the coils directly on the ceiling is usually the simplest and easiest method of installation. Fig. 6-32 shows copper tubing applied to a gypsum lath ceiling. However, copper-tube radiant panels can be applied to any type of lathing. As shown in Fig. 6-33,

251

**Fig. 6-32. A radiant panel installed on a gypsum lath ceiling.**

**Fig. 6-33. The ceiling panel need not be level if the tubing is installed in such a manner as to eliminate pockets that might trap the water.**

the ceiling need not be level *if the tubing is installed in such a way as to eliminate pockets that would trap the water.*

Ceiling panels should not be installed above plywood, composition board, or other insulating types of ceiling material. Such surfaces have an undesirable insulating effect that diminishes the full heat output of the panel.

**Floor Panels.** Radiant panels are installed in the floor in many instances. When this is done, the most satisfactory arrangement is to embed the coils in the concrete floor slab. The best results are obtained when the tubing or pipe is buried at least 2″ below the

252

floor surface, or deeper if a heavy traffic load is anticipated. Allow at least two weeks for the concrete to set before applying heat, and then apply only gradually. Floor covering of tile, terrazzo, sheet flooring, carpeting, or any commercial tile can then be installed. No damage to rugs, varnish, polish, or any other material will result if the water temperature is kept below the prescribed maximum of 85°F.

Fig. 6-34 shows a typical piping diagram for a radiant floor panel. The same care to avoid low places in the coil should be taken as with ceiling-panel coils.

Some loss of heat to the ground from a floor panel laid directly on the ground will be experienced. This will vary with type of construction and fill, but is estimated to be 10% to 20% of the heat given to the room. Heat loss to the outside through the slab edges will usually be greater than the amount lost straight down to the ground. Waterproof insulation that is ½" to 2" thick should be installed along the edges of the slab.

**Wall panels.** Radiant wall panels are not usually employed except to provide supplementary heat where ceiling and/or floor panels do not give the required degree of comfort. Wall panels are also installed in bathrooms in some homes where the presence of more than normal radiant heat is desirable Fig. 6-35 shows a coil installed on a bathroom wall prior to the plastering operation. Low spots that will trap the water must be avoided to permit complete drainage when necessary.

Wall panels installed in areas other than bathrooms need not extend to the floor. They are usually confined to the upper half of the wall to prevent the screening effect on the heat output by furniture placed against or near the wall. Fig. 6-36 shows the installation details for a typical radiant wall panel.

**Panel Testing.** After the coils have been fastened in place and the system has been completed in every respect, a pressure test is made shortly before the concrete is poured or before the plaster is applied. *It is important to again check the installation to make certain that no sags or water traps exist in the coil assembly.*

Two testing methods are usually used to check radiant-heating coils—an *air test* and a *hydraulic test.* For the air test, compressed air at a pressure of at least 100 psi is used. An ordinary dial-type pressure gauge is installed in the line to be tested, along with a

**Fig. 6-34. A typical radiant floor panel.**

shut-off valve on the inlet side of the gauge. Pressure is applied to the system, after which the shut-off valve is closed and the source of air pressure is disconnected. Any leaks in the system will cause the pressure reading to drop continually. The location of the leak can usually be found by listening for the leaking air or by bubbles formed when a soapy solution is applied to the suspected area.

After a satisfactory air test, each coil should undergo a hydraulic pressure test before being covered. The coils should be filled with water, with care being taken to make sure all air is expelled from the system. The water pressure should be increased to 300 psi and the system checked for leaks. Maintain the pressure for at least two hours and again check for a pressure drop. In high buildings, the

Fig. 6-35. Bathroom wall panels may be extended to the floor for concentrated heating.

HEAT TO HEATED ROOM
THROUGH UNINSULATED WALL
EQUALS ABOUT 25% OF OUTPUT

SET DIRECTION OF FLOW SAME AS
MAINS (UP FOR UPFEED, ETC.)

KEEP PANEL AS HIGH
ON WALL AS POSSIBLE

½ TUBE
SPACING

BALANCING AND SHUTOFF
VALVES IN WALL BOX

WALL PANEL DETAILS
SAME AS CEILING

IF BUILT ON OUTSIDE WALL, USE EQUIVALENT
OF 3″ ROCKWOOL OR MORE

**Fig. 6-36. Installation of a typical radiant wall panel.**

test water pressure should exceed the combined static and pump pressure by 100 psi.

**Boilers.** The size of the boiler necessary for a particular installation may be determined in the same manner as that employed for conventional convection hot-water systems. Add the sum of all panel-heating requirements and the piping losses.

Circulating pumps for use with radiant heating should have a somewhat higher head rating than for convector systems of the same capacity. This is because the coil pressure drop is considerably higher than the drop in a radiator or convector.

It is customary to use a conventional boiler layout, such as the one in Fig. 6-37, for small buildings. In this installation, the return water is mixed with hot water from the boiler by means of the flow control or bypass valve. This mixing valve is regulated by the outdoor and indoor thermostats, and if the boiler circulation is fully stopped, this valve will turn the circulating pump off. The temperature of the boiler water is kept constant.

The rooms may be grouped according to exposure, with all panels having the same exposure connected to a common distributing line, the circulating water in which is regulated by individual mixing valves.

In cases where panels of small heat capacity (lag or inertia) are used, such as ceiling panels with small-diameter tubing and close spacing, the bypass valve at the boiler may be regulated by an aquastat, keeping the hot-water temperature constant. The circulation is governed by the room thermostat turning the pump on and off.

With panels of greater inertia, this system is not recommended. Instead, the water temperature should vary inversely with the outside temperature. With such panels, the time lag between the start of the panel heating period and the start of the room heating period is too great. In addition, if the hot water is kept at a constant temperature, the panel stores a great amount of heat on warm days before the room thermostat is affected, causing an excessive rise or "overshoot" in room temperature. For this reason, regulation of the water temperature by both outside and inside thermostats is preferable.

**Controls.** In the design of any heating system, the primary consideration must be the comfort of the occupants regardless of the weather conditions. To preserve and properly utilize the advantages of radiant heating, it is necessary to maintain a moderate coil temperature that is modulated gradually without violent fluctua-

**Fig. 6-37. Piping for a typical small radiant-heating system.**

tions. Thus, it is obvious that intermittent operation cannot fulfill this requirement.

The method of controlling heat by starting and stopping the circulation of the hot water through the coils is simple, but has many serious disadvantages. Chief among these are:

1. Starting and stopping the circulating pump necessarily means a cycling operation—in other words, the system has to reverse itself many times as called for by changing room conditions. This is poor control at best, even in conventional hot-water heating. However, it becomes much worse when applied to radiant heating. In the heavy mass of concrete, or even the lighter mass of plaster, it will not be able to reverse itself quickly, and a considerable lag will follow, which may result in periodic cycles of discomfort.

2. In a radiant-heating system with several circuits, it is difficult to obtain a perfect balance of circuit resistance and heat reserve for all rooms; therefore, when circulation stops, some rooms cool much more rapidly than others. For the same reason, when circulation is restored, some rooms heat up more rapidly than others, with a resultant lack of overall comfort conditions.

3. The most fundamental objection to the on-and-off principle of control with radiant heat is that it allows the surface of the panel to cool and thereby defeats, intermittently, the whole purpose of radiant heat. As the panel surfaces are allowed to become cold at intervals, the system ceases to function as radiant heating, and the human body then becomes exposed to cold surfaces with a falling air temperature.

Fig. 6-38 shows the actual recorded air temperatures and surface temperatures of a concrete floor in which pipes were embedded and controlled by an on-and-off switch operating the pump according to room temperature. The control was set to start the pump when the air temperature dropped to 68°F. In this case, the bottoms of the pipes were 2″ below the top surface of the concrete floor. The lag and the amount of variation will, however, vary according to the depth of the pipes below the surface and the type of floor covering used.

It has been proved by many years' experience that a well-

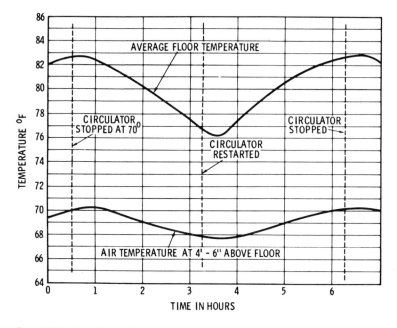

Fig. 6-38. Variation of air and surface temperature with an on-and-off control.

controlled continuous-flow system is much more economical than an on-and-off system. Obviously then, the proper way of controlling radiant panels is to allow continuous circulation of water in the coils at all times. The water temperature should be modulated according to outside temperature changes, and an indoor thermostat sensitive to radiant heat should be used as a limit control to prevent overheating without stopping circulation.

A typical automatic water blender controlled by the outside air temperature is shown in Fig. 6-39. A typical boiler installation using this type of blender is illustrated in Fig. 6-40.

The water blender consists principally of a thermostatic three-way valve adapted to recirculate a variable portion of the return water, automatically adding the proper amount of hot water from the boiler to maintain the correct coil temperature. A simple liquid expansion thermostat is mounted on an outside wall of the building, preferably with a northwest exposure and away from the direct rays of the sun. Pressure built up in this thermostat as it

FLEXIBLE TUBING

K

ADJUSTMENT

LIMIT STOP

HOT WATER
THERMOMETER

LOCKNUT

J

3

TO SYSTEM

OUTSIDE
THERMOSTAT
ELEMENT

INSIDE THERMOSTAT
ELEMENT

A

FACTORY SET
ADJUSTMENT

B

FROM SYSTEM

2

S

FROM
BOILER

1

Courtesy Sarcatherm Controls, Inc.

**Fig. 6-39. Construction of an automatic water-blender valve controlled by
an outside thermostatic element.**

Fig. 6-40. A typical installation using an automatic water-blender valve controlled by an outside thermostat.

absorbs heat is transmitted through a flexible tube to element $A$ in the valve, pushing piston $B$ downward. Hot water enters the valve at $1$, and return water from the system through inlet $2$. The hot and cooler return water pass through the mixing valve $S$, then over element $A$, and to the system through outlet $3$.

Element $A$ forms a part of the thermostatic system, so that the position of mixing valve $S$ is determined by the combined influence of the temperature at the outside thermostat and temperature of the mixed water flowing over $A$. The two elements ($A$ and $S$) of the thermostatic system are so proportioned as to supply the heating coils with water at exactly the temperature needed for the existing weather conditions.

Should it be necessary to raise or lower the water temperature, the desired adjustment can be made by loosening locknut $J$ and turning knob $K$. All controls are calibrated at the factory for the particular needs of the system.

Heat loss from buildings is subject to several variables, not all of

261

which can be adequately controlled in accordance with the fluctuations of the outside temperature alone. Exposure, wind velocity and direction, as well as the insulating properties of the walls, all exert their influence. Because there is a certain time lag between changes in atmospheric conditions and the temperature in the rooms, control by outside temperature alone may result in rooms becoming too warm at times. An automatic check against this overheating can be provided by a compensating indoor control.

This control system is similar to that shown in Fig.6-39 and 6-40, except that an additional temperature-sensitive bulb $Q$ (Fig. 6-41) is added. This bulb is fitted with a small electric heater coil $(H)$ and a manually adjustable resistance $(R)$ mounted in a metal box for convenient attachment to the wall in the boiler room. A thermostat is placed in a key room from which overheating by the system can be controlled. The thermostat has a normally open switch that acts to close the circuit to $H$ when the desired room temperature is reached. This action applies heat to bulb $Q$, which will cause a greater pressure in the thermostatic flexible tube system, closing the hot-water inlet port and recirculating the cooler return water through the heating coils. This effectively checks the overheating while preserving the responsiveness of the system to outside temperature changes. The piping diagram of a typical system using this indoor-outdoor control feature is shown in Fig. 6-42.

An automatic heat-control system for use with heavy radiant-heat panels is shown in Fig. 6-43. In this system, the outdoor controller senses any change in the heating load as the outdoor temperature varies. This change is transmitted to the control panel, where a corresponding change is made in the setting of a temperature controller whose sensing bulb is located in the supply water line to the heating coils.

If a thermostat is used, it senses changes in the internal heat load caused by occupancy, solar gain, and so on, and acts as a limit control to prevent overheating. The circulator operates continuously except when no heat is required. The safety control prevents excessive temperature of the boiler water.

The automatic heat-control system in Fig. 6-44 is for use with light radiant-heat panels. This type has no outdoor sensing unit. Instead, the room thermostat senses any change in room temperature and transmits this change to the control panel, which adjusts

OUTDOOR BULB

INDOOR ROOM CONTROL-
NORMALLY OPEN SWITCH

FLEXIBLE TUBING

THERMOMETER

FIXED
RESISTOR

CONTROL BOX

③
TO SYSTEM

R

Q

INDOOR
AUXILIARY
BULB

H

HEATING COIL

FROM
SYSTEM

②

GROUND

110-VOLT
LINE

SWITCH   FUSE

FROM
BOILER

①

Courtesy Sarcatherm Controls, Inc.

**Fig. 6-41. An automatic water-blender valve with indoor compensation
feature.**

the control point of a temperature controller whose sensing bulb is
located in the supply water to the heating coil. The temperature
controller repositions the three-way valve, which mixes the hot
boiler water with the cooler return water from the bypass to

263

**Fig. 6-42. Heating system using indoor-outdoor controls.**

**Fig. 6-43. An automatic heat-control system for heavy radiant-heat panels.**

provide a mixture of the proper temperature to satisfy the setting of the temperature controller. The burner is controlled by an aquastat (immersion temperature controller) to maintain the necessary temperature of the water in the boiler for the indirect water heater. The circulator is shut down by the auxiliary switch whenever the three-way valve is positioned for zero heating. The safety control will shut off the burner if, for any reason, the water temperature in the boiler reaches the maximum safe level.

Courtesy Honeywell, Inc.

**Fig. 6-44. An automatic heat-control system for light radiant-heat panels.**

Courtesy Honeywell, Inc.

**Fig. 6-45. A typical control system for a multiple-zone radiant-heating system.**

A multiple-zone-control radiant-heating system is shown in Fig. 6-45. The outdoor controller measures the change in the heating load caused by a change in the outdoor temperature. This change is transmitted to the control panel, causing a corresponding change in the setting of a temperature controller whose sensing bulb is located in the main distribution line, and downstream from the three-way valve. The temperature controller positions the three-way valve to mix the hot boiler water with the cooler return water in the proper proportions to furnish supply water to the heating coils at the proper temperature. A temperature controller in the boiler controls the burner so as to maintain the boiler water at the temperature required by the indirect heater. A room thermostat in each zone controls a motorized valve in its particular supply line. This valve regulates the panel-coil temperature in accordance with the individual requirements of the zone.

CHAPTER 7

# Boiler Fittings

Certain fittings or devices usually mounted or connected to the boiler in steam and hot-water heating systems are essential to the proper operation of the boiler. Among these are:

1. Valves
    a. Safety.
    b. Stop.
    c. Check.
    d. Blow-off.
2. Water-gauge cocks.
3. Water gauges.
4. Steam gauges.
5. Injectors.
6. Fusible plugs.

All of these fittings will not necessarily be installed on all boilers (especially hot-water boilers), but the pipe fitter should be familiar with their purpose. Fig. 7-1 shows the piping diagram of a typical steam boiler with the control and indicating devices essential to its proper operation.

**Fig. 7-1. A typical boiler showing the control and indicating devices essential to its proper and safe operation.**

# VALVES

The valves used in boiler installations are of several types including the familiar globe and gate valves. Only the best-quality valves should be used because of the demands placed on them by the exposure of their working parts to the high temperatures of the steam or hot water they are controlling.

## Safety Valves

Every boiler installation must have a *safety valve* incorporated to protect the boiler itself and the building occupants in case of malfunction. These valves are adjusted to open and relieve the internal pressure, should it rise above a safe predetermined level. A boiler water safety relief valve is shown in Fig. 7-2.

**Fig. 7-2. A boiler-water safety relief valve.**

Courtesy Watts Regulator Co.

## Combination Feeder and Low-Water Cutoff

When a combination low-water feeder and cutoff is used on a steam boiler, the feeder adds make-up water when needed for normal operation. If, for any reason, an emergency occurs, the low-water cutoff operates to shut down the firing mechanism by interrupting the electrical current to the burner. When the emergency has passed, the water feeder resumes normal operation. The combination feeder shown in Fig. 7-3 has fittings for a gauge-glass connection.

## Low-Water Cutoff

A low-water cutoff (Fig. 7-4) automatically shuts off the burner

Fig. 7-3. A combination boiler-water feeder and low-water cutoff.

Fig. 7-4. A boiler low-water cut-off.

or stoker whenever the water level in the boiler falls below a safe operating minimum. The cutoff is mounted with the float level with the boiler water level. If the water level drops, the float also drops, breaking the electrical circuit to the burner or stoker.

## Stop Valves

Globe or gate valves are commonly used as *stop valves*. They are merely valves placed in pipe lines—either steam, hot water, or cold water—to allow the passage of steam or water, or to block it completely. They normally are turned completely on or completely off, seldom being used to regulate the flow in any intermediate manner.

## Stop-Check Valves

Stop-check valves should not be confused with ordinary stop valves. A stop-check valve is primarily designed to act as an automatic nonreturn valve by preventing a backflow of steam from the main steam header to the boiler to which it is connected, in the event of failure of that boiler. Stop-check valves are used when more than one boiler is connected to a common steam header.

## Check Valves

It is sometimes necessary to allow steam or water to flow through a pipe in one direction but not in the other. An example of this is in the cold-water supply line to the boiler. A *check valve* is normally installed in this line to permit water to enter the boiler but prevent water (or steam) in the boiler from backing up into the supply line. These valves usually have a swinging gate and must be installed horizontally with the hinge at the top of the valve.

## Blow-off Valves

A valve installed in a line connected to the lowest part of the water chamber in a boiler is called a *blow-off valve*. It is used to drain off sediment, scale, and other impurities that settle out of the water as it is heated. This settlement accumulates at the lowest point in the boiler, where it is removed by periodic opening of the blow-off valve.

# BOILER WATER LEVELS

It is very important that the water level in the boiler be maintained within certain limits at all times. If the level is too high,

271

inefficient operation results—if too low, there is danger of permanent damage to the boiler or the possibility of an explosion. Two principal methods are used to determine the water level.

## Try Cocks

Small valves called *try cocks* are installed on steam boilers at the safe high-water level and at the safe low-water level. On larger boilers, a third try cock, halfway between the other two, is sometimes used. Fig. 7-5 shows a typical handwheel type of try cock. For large steam boilers where the try cocks are at a level so high as to be unhandy to reach, a lever (Fig. 7-6) is used which is opened and closed by means of a chain.

Fig. 7-5. A bronze handwheel try cock.

Courtesy Eugene Ernst Products Co.

Fig. 7-6. A bronze lever-type try cock.

Courtesy Eugene Ernst Products Co.

To ascertain the water level, first one cock and then the other is opened slightly, and the presence of water or steam is determined by its appearance or sound. With a little experience it becomes easy to determine the approximate water level.

The correct method must be used to check the water level with try cocks. Fig. 7-7 shows the results of the right and the wrong way when the water is at a particular level with respect to the try cock. By opening the try cock slightly, the escape of steam indicates the water level is below the fitting. If, however, the level is only slightly below the try cock and the cock is opened wide, the resulting violent agitation of the water will cause some of it to escape, giving the false impression that the water level is at or above the fitting. Therefore, try cocks should be opened only slightly to ensure that a false water level is not indicated.

**Fig. 7-7. Correct and incorrect methods of testing the water level in a steam boiler.**

## Water Gauges

Water gauges are installed in steam boilers to give a visual indication of the boiler water level. When a low-water cutoff is used as shown in Fig. 7-3, the gauge glass can be mounted on the low-water cutoff.

273

A water gauge (Fig. 7-8) consists of a strong glass tube long enough to cover the safe range of water level and having the ends connected to the boiler interior by fittings. It is connected to the boiler at such an elevation that the lower and upper ends of the tube will be below the lowest and above the highest permissible water level. The internal construction of a typical water gauge is shown in Fig. 7-9.

**Fig. 7-8. A typical two-rod plain water gauge.**

Courtesy Lunkenheimer Co.

## Water Columns

Try cocks and water gauges are frequently connected to a special fitting known as a *water column* (Fig. 7-10). This device contains floats that actuate a whistle when the water level drops below a safe level. This alerts the boiler operator to add water. It is recommended that local code requirements be checked before a water column is installed.

Provision is made for installing crosses with plugs at all right-angle turns, since the pipes and fittings often stop up with scale and sediment. After the steam is up, make sure the column connec-

**UPPER BODY**    **PLUG**    **BONNET RING**    **BONNET**   **GLAND**    **PACKING NUT**

**HANDWHEEL**

ROD PLATE

GLASS STUFFING BOX

GLASS PACKING

GLASS STUFFING-
BOX WASHER

GLASS

ROD

STEM    PACKING    HANDWHEEL
LOCKNUT

LOWER BODY

DRAIN-UNION TAIL

DRAIN-UNION RING

Courtesy Lunkenheimer Co.

**Fig. 7-9. Cross-sectional view of a typical water gauge.**

275

Courtesy Eugene Ernest Products Co.

Fig. 7-10. A water column with a vertical gauge.

tions are unobstructed by observing the action of the water in the gauge glass. Blow down the column (opening the valves fully for a short period of time) for a double check, which will also test the low-water alarm. Keep the gauge glass clean and properly lighted for good visibility. Operate the gauge cocks regularly to ensure their proper operation for emergency service.

## Sight-Glass Maintenance

Good maintenance pays dividends in safety and troublefree operations. A few pointers for sight-glass maintenance are:

1. Keep sight glass in its original container until ready to install.
2. Avoid bumping, chipping, or scratching.
3. Gasket seats must be flat, smooth, and free of metal of gasket bits.
4. Only approved gaskets should be used. These should be at least $\frac{1}{16}$", but not more than $\frac{1}{4}$" thick on the vessel sides. Clampring gaskets should be less than $\frac{1}{16}$" thick. They should be of smooth material without wire inserts.
5. All bolts and nuts should be free of burrs, and should be well lubricated with graphite grease. They should be tightened to the value specified by the manufacturer (usually 10 to 20 lb.-ft., depending on the number of bolts and their size). A regular tightening sequence should be followed to avoid uneven loads on the glass. The use of conical spring washers will keep the joints tight during cycling.
6. A commercial glass cleaner should be used to keep the glass clean. Avoid the use of wire brushes, scrapers, and harsh abrasives. Inspect and replace the glass tubes on a regular schedule.
7. Use a bright light to check for glistening marks in or on the glass, and for cloudy or roughened surfaces. If any of these signs are found, replace the glass regardless of the maintenance schedule.
8. A replacement glass must have a pressure and temperature rating higher than the service rating of the vessel. Be sure to replace all shields intended to protect the glass from falling objects of sudden blasts of cold air.

It is sometimes necessary or desirable to cut a sight glass from a long section of tube. This can be accomplished quite easily if a sight-glass cutter, such as the one in Fig. 7-11, is used. Follow these steps:

1. Score the inner surface of the tube at the correct point. It is important to score once around only—additional scoring will result in a jagged edge and will dull the cutter.
2. Heat the outside of the glass tube directly over the score for a few seconds with the tip of the hot blue flame of a Bunsen burner, propane torch, or oxyacetylene flame; the glass will crack audibly and visibly at the score. Heat only until the crack encircles the glass.
3. If only a short piece is to be cut off, tap the scored end lightly on a hard surface, preferably wood. If a long piece is to be cut off, grip the glass firmly on each side of the score and pull apart without bending.

**Fig. 7-11. A boiler sight-glass cutter.**

## STEAM GAUGES

A steam gauge is a very important fixture that should be tested at frequent intervals to ascertain if it is indicating the correct steam pressure. This device indicates the difference in pressure between the inside and outside of the boiler. In other words, it indicates a *gauge pressure* as distinguished from *absolute pressure*.

A steam gauge works on one of two principles—the expansion of a corrugated diaphragm when pressure is applied, or the tendency of a curved tube to assume a straight position when under pressure.

The bent-tube gauge is now almost universally used. Since the movement of the free end is necessarily small, its motion is multiplied by means of a rack-and-pinion gear, which causes a pointer to rotate. The principle of operation is shown in Fig. 7-12.

When a gauge is in good working order, the pointer will move easily with every change of pressure in the boiler. The pointer should return to zero when the steam is shut off.

A properly installed gauge should have one or more turns of tubing placed between it and the boiler, as in Fig. 7-13. This "pigtail" fills with condensate to prevent live steam from reaching the working parts of the gauge, thus preventing damage. The gauge will be ruined if such a "pigtail" is not used, especially if superheated steam is produced. In fact, before installing a gauge, *the "pigtail" should be filled with water*. Locate the gauge in a cool place and secure it to some substantial object where it will be free from vibration or shock.

**Fig. 7-12. Mechanism of a bent-tube steam gauge.**

## INJECTORS

An *injector* is a device for forcing water into a boiler against the boiler pressure by means of a steam jet. The simplest form of an injector is shown in Fig. 7-14. Steam from the boiler enters the fixed steam nozzle and passes through the space between the

**Fig. 7-13. Steam-gauge pigtail.**

nozzle and combining tube. It enters the combining tube and then passes into the overflow chamber and out through the overflow check valve. As it passes from the steam nozzle into the combining tube, a vacuum is formed that draws water in through the water inlet. The inrush of cold water condenses the steam in the combining tube, and the water thus formed is at first driven out through the overflow. As the velocity of the water jet increases, however,

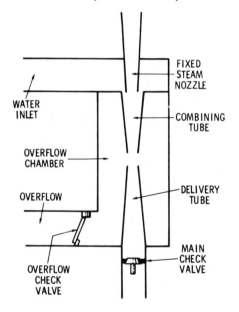

**Fig. 7-14. Elementary principles of an injector.**

sufficient momentum is obtained to overcome the boiler pressure; water enters the delivery tube, passes the main check valve, and enters the boiler. With the water now flowing into the boiler, a vacuum is produced in the overflow chamber; the vacuum causes the overflow check valve to close, preventing air from entering the delivery tube. This sequence of events is shown in Fig. 7-15.

**Fig. 7-15. Starting-cycle sequence of an injector.**

There are three pipe connections to be made to an injector. One is the steam pipe, which must be the same size as the injector connection. This pipe must be connected to the highest possible point on the boiler and be independent of any other pipe in order to secure the best results. It must be blown out with the steam before the final connection to the injector is made.

The second pipe connected to the injector is the suction pipe. It must be at least as large as the injector connection. Where the lift is over 10′, the suction pipe should be one or two sizes larger, reduced to the injector size as close as possible to the injector and with a glove valve the same size as the pipe installed as near the injector as possible. A second globe valve is sometimes installed in this line if the water source has a high pressure. It is used to reduce the pressure while the first valve regulates the amount of water.

A foot valve should be installed on the low end of the suction pipe where a long lift is necessary. Without a foot valve, the air in the pipe must be exhausted each time water is called for, resulting in a considerable waste of steam.

The delivery pipe should have both a globe valve and a check valve installed between the injector and the boiler. A drain pipe to take care of the overflow completes the installation. A cross-sectional view of a typical injector is shown in Fig. 7-16.

**Fig. 7-16. Cross-sectional view of a typical injector.**

## FUSIBLE PLUGS

A *fusible plug* is a safety device installed on some boilers to protect it in case of dangerously low water. These plugs (Fig. 7-17) are commonly made of bronze and filled with pure tin. The melting temperature of the tin is approximately 450°F, at which

(A) Inside type.

(B) Outside type.

Courtesy Lunkenheimer Co.

**Fig. 7-17. Fusible plugs.**

point the plug will relieve the pressure within the boiler. Fusible plugs are normally used only on boilers having a steam pressure of less than 250 psi.

## PRESSURE-RELIEF VALVES

Steam and hot-water boilers must be equipped with an approved *relief valve* installed in accordance with local or ASME codes. Such a valve is shown in cross section in Fig. 7-18. When pressure within the boiler reaches a certain level, the relief-valve mechanism actuates to relieve that pressure. After the pressure is reduced to a safe level, the valve closes.

Fig. 7-18. Typical water-pressure-
relief valve.

## STEAM LOOP

An arrangement of piping called a *steam loop* is used in a steam system to return condensate to the boiler. The essential parts of a steam loop are shown in Fig. 7-19. The *riser* contains a mixture of water and steam. The steam portion of the mixture is condensed by means of the *condenser* at the top. This condensation reduces the pressure in the system, and the reduced pressure causes the mixture in the riser to flow upward.

As soon as the water mixed with steam passes the *goose neck*, it cannot return to the riser. Hence, the contents of the pipes constantly work from the separator toward the boiler, the condenser being slightly inclined toward the *drop leg*. The condensate will accumulate in the drop leg to a height such that its weight will balance the weight of the mixture in the riser.

# HEATING PUMPS

There are two types of pumps designed for use in steam heating systems—the *condensation pump*, which is used with a two-pipe gravity heating system, and the *vacuum pump*, which is used with return-line vacuum and variable-vacuum heating systems. Although the design and ultimate operation of these two types differ, they have one fundamental feature common to both—the accumulation of the condensate and its delivery to the boiler. The water condensed from the steam supplied to the steam heating system is the source of condensate.

Wherever the condensate is not wasted to the sewer, the cycle through which the medium passes in steam heating systems is as follows:

1. Vaporization into steam.
2. Conveyance to the radiation.
3. Condensation from steam to water.
4. Conveying condensation to the pump.
5. Return of the condensate to the boiler.

Fig. 7-19. Essential parts of a steam loop.

285

Where the condensate is lost from the system or is wasted to the sewer, the fourth and fifth phases of the cycle may include wasting the condensate and supplying water from a source of "make-up."

Pumps should be used for steam heating systems if any of the following conditions exist:

1. If the pressure at the boiler is too great to permit the condensate to return to the boiler by gravity or by a boiler return trap.
2. If the dry-return mains are below the boiler water line.
3. If the size of the heating system, in terms of condensing capacity, is beyond the capacity of the largest return trap.
4. If the condensate must be forced from a lower elevation over a "high-point" or to the floor level of the boiler room; for example, where several buildings are fed from one boiler plant and the condensate has to be raised from the basement of one and forced through an underground return to a higher level to reach the boiler room.

## Condensation Pumps

Condensation pumps are available as complete, compact assemblies (packages) for returning water to low- and medium-pressure boilers in gravity heating systems, steam process equipment, or combinations of both; and are available as components that may be assembled (installed) on the job to fit special conditions. They are also used in low-pressure systems where the return mains are located at elevations that do not permit a gravity flow of the condensate to the boiler.

These pumps are operated in response to a float switch and will quickly elevate the condensate from a low to a high return line, or will automatically pump the condensate from auxiliary apparatus to storage tanks. Their use permits the increase of usable heated space in a building by allowing the radiation to be installed at a level lower than the water level in the boiler.

These pumps are available as a single pump unit (Fig. 7-20) or a duplex (Fig. 7-21). The duplex unit comprises two pumps with a receiver to which the suction of each pump connects. Duplex pumps usually employ an alternator, which transfers the operation from one pump to the other, in sequence, as the water level varies

Courtesy Dunham-Bush, Inc.

Fig. 7-20. A heavy-duty, single-unit condensation pump.

Courtesy Dunham-Bush, Inc.

Fig. 7-21. A duplex condensation pump.

between the cut-in and cut-out points. In addition, they provide automatic standby service so that if one pump fails to start or cannot handle the load, the second pump starts automatically. Fig. 7-22 illustrates a typical piping and wiring arrangement for controlling a single condensation pump to maintain a uniform water level in a single boiler.

287

## Vacuum Pumps

Vacuum pumps are similar to condensation pumps in one of their functions—they receive system condensate and pump it to the boiler. However, a vacuum pump has an added function. It produces a vacuum by removing air, vapor, and condensate from the system. The vacuum thus created causes circulation in the heating system before the pressure has been raised in the boiler. By quickly exhausting air (noncondensate gases) and condensate from the system, the vacuum pump induces the steam to circulate rapidly, reducing the warmup time and allowing the heating system to function quietly. Fig. 7-23 shows a typical vacuum pump. Vacuum pumps are rated so that each standard unit will easily handle the requirements of any heating system up to the capacity of the pump without special allowance for covered piping.

Duplex units have the advantage over single pumps due to the provision of automatic standby service. Should one pump fail to start, or be unable to handle the load, the other pump is cut in and

Courtesy Dunham-Bush, Inc.

Fig. 7-22. Typical piping and wiring arrangement for controlling a single condensation pump to maintain a uniform water level in a single boiler.

COPPER FLOAT

VENT

AIR SEPARATING TANK

JET EXHAUSTER

VACUUM SWITCH

FLOAT SWITCH

COPPER FLOAT IN ACCUMULATOR TANK

CENTRIFUGAL PUMP

RETURN PIPING CONNECTION TO ACCUMULATOR TANK

DISCHARGE LINE

DISCHARGE VALVE

**Fig. 7-23. A single-unit vacuum pump.**

Courtesy Dunham-Bush, Inc.

picks up the load since it is controlled by vacuum or float switches independent of the other pump.

## Pump Selection

Whether a condensation pump or a vacuum pump is necessary, in most instances, depends on whether a lowest first cost or the best investment and maximum system performance is the major consideration. Some of the advantages of the vacuum pump over the condensation pump are as follows:

1. Rapid heating, when the system first starts up, is achieved by the addition of a vacuum pump. Air is evacuated from the system ahead of the steam so that the steam is distributed while pressures are still low. Circulation to all radiators is assured because a vacuum in the return line continually removes any air released by the thermostatic traps. This results in savings in fuel costs because the boiler will operate under vacuum in mild weather.

2. If the return lines are too low to reach the pump inlet by gravity, a separate accumulator tank may be placed at a lower level than the pump and the condensate raised from it by the vacuum the pump produces. Thus, in some installations, the need for a large pump pit may be eliminated.

3. Pipe sizes may be reduced in systems that have vacuum pumps. This results in substantial savings on installation costs. The cost of a vacuum pump is greater than that of a condensation pump, but the advantages of the vacuum pump usually will offset this cost.

# CHAPTER 8

# Fuel-Oil Tank Installation

Fuel-oil tanks must conform in construction and installation to the regulations of the National Board of Fire Underwriters and to local ordinances. Vents, fill pipes, outlet pipes, and gauges are also governed by the regulations. Fuel-oil tanks may be classified with respect to their location, as:

1. Gravity tanks.
2. Subgravity tanks.
3. Inside tanks (basement).
4. Outside tanks (underground).

They may also be classified with respect to the number of pipe connections between the fuel tank and the oil burner, as:

5. Single-pipe installations.
6. Two-pipe installations.

Gravity tanks, which are the most common, are those placed at a higher level than the burner so that the fuel is delivered by gravity to the burner. Fig. 8-1 shows a typical gravity installation of a tank installed outside the building.

**Fig. 8-1. A typical underground, two-pipe tank installation with the oil burner below the level of the tank unit.**

Subgravity tanks are those from which the fuel is drawn by a pump, which delivers it to the burner. Most burner pumps in installations of this type will lift the oil 12′ or more while delivering it to the burner at the required pressure. Fig. 8-2 shows a typical subgravity installation of a tank installed outside the building.

Single-pipe installations are those in which only a suction line is required between the oil tank and the burner; whereas in two-pipe installations, two fuel lines are employed, one a suction line and the other an oil-return line.

A single-pipe installation, such as shown in Fig. 8-3, is used only where the bottom of the oil tank is above the burner pump. If there is any oil in the tank, there will then be a gravity flow to the pump. The suction or supply line is connected to the bottom end of the tank. The oil pumped in excess of that used is recirculated or bypassed at the burner, from the pressure-valve discharge into the strainer or filter housing on the inlet side of the pump.

**Fig. 8-2.** A typical underground, two-pipe tank installation with the oil burner above the level of the tank unit.

**Fig. 8-3.** An inside, single-pipe tank installation with the oil burner below the level of the tank unit.

293

A two-pipe or suction-and-return method of installation is used on inside tanks when the bottom of the tank cannot be set above the pump and on all outside underground tank installations.

## TANK CAPACITY

If space is available in the basement, a 275-gal. inside oil storage tank is generally preferred. The customary dimensions of such a vertical-oval tank are approximately 27″ by 42″ on the oval end, by about 64″ in length. Oil supply tanks located inside buildings should not exceed 550 gals. individual capacity, or 550 gals. aggregate capacity.

A typical piping connection diagram for a double inside tank installation is given in Fig. 8-4. In locations where available basement space is at a minimum, or where it is desirable to store larger quantities of oil, an underground tank outside the building is used. Outside tanks are usually of 1000-gal. capacity or more.

## INSTALLATION

Where there are no regulations governing the installation of tank and burner, the following recommendations should be followed.

1. In selecting a location for the tank, whether in the basement (Fig. 8-5) or buried outside (Fig. 8-6), keep in mind the accessibility for filling, and locate the tank the proper distance from the heating plant. Inside tanks should be at least 7′ from the heating plant, should be properly vented to the outside of the building, and that the fill line should be installed from the outside. The vent line should not be smaller than 1 ¼″, and the fill line should be from 1 ¼″ to 3″. Outside tanks should be buried below the frost line and connected with the proper fill and vent pipes as well as the suction and return lines to the burner.

2. The suction line, where possible, should be ⅜″ O.D. copper tubing. One-half-inch O.D. copper tubing should be used for the return line. Make sure there are no leaky connections

**Fig. 8-4. A method of piping a double-tank installation.**

between the tank and the fuel unit. Continuous lengths of tubing should be used for the return and suction lines.

3. It is always advisable to run the return line to the same depth in the tank as the suction line on outside tank installations. This will avoid the possibility of air getting into the pump during the shutdown period of the burner.

295

Fig. 8-5. An inside-tank installation showing the piping and other details.

**Fig. 8-6.** An outside-tank installation showing the piping and other details.

4. Keep in mind that when a return line is used, a bypass plug must be inserted in a Sundstrand pump is used, or the top cover must be reversed if a Webster pump is used.

5. Whenever the top of the tank is above the burner level, an approved make of siphon-breaking or antisiphon valve should be installed in the suction line. The purpose of this valve is to prevent the contents of the tank from siphoning into the basement, should a pipeline break. A horizontal swing-check valve should be installed in the suction line at the point where the line enters the basement.

It is recommended that inside storage tanks be provided with a drain opening. When such a drain opening is provided, the tank should be installed with the bottom slanted toward the drain opening, with a slope or not less than ¼ " per foot of length. The drain opening should be provided with a suitable pipe connection in a form to provide a sump from which water or sediment can be readily drained at intervals.

Prior to setting an underground tank, it should be protected against erosion on the outside in a manner satisfactory to the inspection department having jurisdiction, but in every case at least equivalent to two primary coatings of red lead followed by a heavy coating of hot asphalt. Inside installations and auxiliary tanks should be securely supported by a substantial incombustible support to prevent settling and sliding.

Outside above-ground tanks should be set on a firm foundation at least 1' above ground level, and should have supports of masonry or protected steel, except that wood cushions may be used. No combustible material should be stored under or within 10' of outside above-ground tanks.

Underground tanks and tanks inside buildings shall be constructed of steel or wrought iron of a minimum gauge (U.S. Standard) in accordance with Underwriter's specifications, and with local ordinances, if any.

## Tank Vents

Storage tanks should be equipped with an open vent or an approved automatically operated vent arranged to discharge to the open air. Openings and vent pipe should be of ample size to prevent abnormal pressures in the tank during filling, but not smaller than 1½".

Where storage tanks are filled by the use of a pump through tight connections, special consideration should be given to the size of the vent pipe to ensure that it is adequate to prevent the development of abnormal pressure in the tank during filling. This may be accomplished by providing a vent pipe commensurate in size with the discharge of the pump used or by means of an approved signaling or limiting device.

The vent pipe should be arranged to drain to the tank. The lower end of the vent pipe should not extend through the top into the tank for a distance of more than 1". The vent pipe should terminate outside the building at a point not less than 2', measured vertically or horizontally, from any window or other building or opening. The outer end of the pipe should be provided with a weatherproof hood and should terminate far enough above the ground to prevent its being obstructed with snow and ice.

The vent pipe should not be cross-connected with the fill pipe or return line from the burner. The vent opening on outside aboveground tanks should be provided with approved flame arresters. Local building regulations must be followed when installing the oil-tank vent pipe. Some areas require that an oil-tank vent be extended to or above the roof line.

### Fill and Overflow Pipes

Storage tanks, other than outside above-ground tanks, should be filled only through fill pipes terminating outside the building at a point at least 5' from any building opening at the same or lower level. When not in use, fill terminals should be closed tightly by a metal cover designed to prevent tampering. Auxiliary tanks should be filled by pumping from storage tanks. Cross-connections permitting gravity flow from one tank to another should not be made. This, however, should not be construed as prohibiting filling of an outside tank by gravity.

All fill pipes should be located near a driveway or street curb to avoid the necessity of walking on soft or soggy lawns with a fill hose during spring and fall. In the event that a furnace is to be installed during cold weather, it is desirable to have the tank installed and fuel delivered before starting to install the burner.

## Valves

A readily accessible shut-off valve of an approved type should be installed in the oil supply line near the burner. A shut-off valve of an approved type should be installed on each side of an oil strainer that is not part of the oil burner unit, and on the discharge and suction line of an oil pump directly to the burner, but which is not part of the burner unit.

# TESTING

After installation and before being covered, tanks and piping should be tested hydrostatically or with air, at a pressure from 5 to 10 pounds per square inch, but not less than 5 pounds per square inch at the highest point of the system. Instead of a pressure test, suction lines may be tested under a vacuum of not less than 20″ of mercury. Tests should continue for at least 30 minutes without a noticeable drop in pressure or vacuum.

CHAPTER 9

# Gas Piping

The word "piping," as used in reference to the installation of gas appliances, refers only to rigid conduit of steel, iron, brass, or copper that is acceptable under local codes. Installers should also refer to the manufacturer's instructions, gas supplier regulations, and local building, heating, plumbing, or other codes in effect in the area in which the installation is made.

Installation and replacement of either gas piping or gas appliances should be done only by a qualified installer; one who knows the laws and rules involved, as well as the necessary safety precautions.

## GENERAL PRECAUTIONS

When installing gas piping or gas appliances, all gas should be turned off in order to eliminate leakage hazards. Before turning off the gas to the premises for the purpose of installation, repair, replacement, or maintenance of gas piping or appliances, all burners should be turned off. When two or more consumers are served from the same supply system, measures should be taken to assure that only service to the proper consumer is turned off.

When checking for gas leaks, only a soap-and-water solution, or other material accepted for that purpose, should be used. Matches, candles, flames or other sources of ignition should not be used for this purpose. Artificial illumination should be restricted to battery-powered flashlights (preferably of the safety type) or approved safety lamps. In searching for leaks, electric switches should not be operated. If electric lights are already on, they should not be turned off. Smoking should not be permitted.

## INSTALLATION OF GAS PIPING

Before proceeding with the installation, a piping sketch or piping plan should be made of the gas piping system. This plan should show the proposed location of the piping as well as the size of the various branches. Adequate provisions should be made for additional appliances that might be installed later.

Prior to piping installation, however, the piping layout or specifications should be submitted for approval to the gas supplier or the authorities having jurisdiction. When additional appliances are to be served through any existing piping system, the capacity of the system in service should be checked for adequacy and replaced with larger piping if necessary.

### Location of the Gas Meter

Meters should be located as near as practicable to the point of entrance of the gas service and where the connections are readily accessible for servicing. Location, dimensions, space requirements, and type of installation should be acceptable to the gas supplier.

Gas piping should be sized and installed to provide an uninterrupted supply of gas sufficient to meet the maximum demand without undue loss of pressure between the meter (or service regulator when a meter is not provided) and the appliance or appliances. The size depends on the following:

1. Allowable loss in pressure from the meter (or a service regulator when a meter is not provided) to the appliance.
2. Maximum gas consumption to be provided.
3. Length of piping and number of fittings.
4. Specific gravity of the gas.
5. Diversity factor.

The specific gravity of most gases varies from 0.45 to 0.65, and since the capacity of the piping is affected by this factor inversely as the square root, it is usually sufficiently accurate to use an assumed gravity factor of 0.60 for all calculations.

## Gas Consumption

The quantity of gas to be provided at each outlet should be determined directly from the manufacturer's Btu rating of the appliance to be installed. If the ratings of the appliances to be installed are not known, Table 9-1 lists the approximate consumption of the average appliance in Btu per hour.

## Size of Piping

Capacities for different sizes and lengths of pipe in cubic feet per hour for gas of 0.60 specific gravity, based on a pressure drop of 0.3″ of water column are shown in Table 9-2. In adopting a 0.3″ pressure drop, due allowance for an ordinary number of fittings should be made. Table 9-2 is based on a gas of 0.60 specific gravity, but if it is desired to use an exact specific gravity for a particular condition, the values in the table can be corrected by multiplying by $\sqrt[\circ]{0.6/\text{sp.gr.}}$

To obtain the size of piping required for a certain unit, it is first necessary to determine the number of cubic feet of gas per hour consumed by the unit as follows:

$$\text{Cu. ft. of gas/hr.} = \frac{\text{total Btu/hr required by unit}}{\text{Btu/cu. ft. of gas}}$$

*Example:* It is required to determine the size of the piping in a gas-fired boiler installation when the burner input is 155,000

## Table 9-1. Approximate Gas Consumption of Typical Appliances

| Appliance | Input Btu/hr (approx.) | | |
|---|---|---|---|
| Boiler or furnace (domestic)............. | 100,000 | to | 250,000 |
| Range (freestanding, domestic).......... | 65,000 | | |
| Built-in oven or broiler unit (domestic) ... | 25,000 | | |
| Built-in top unit (domestic).............. | 40,000 | | |
| Water heater, automatic storage | | | |
| (50-gal. tank)....................... | 55,000 | | |
| Water heater, automatic instantaneous | | | |
| 2 gal. per minute.................... | 142,800 | | |
| 4 gal. per minute.................... | 285,000 | | |
| 6 gal. per minute.................... | 428,400 | | |
| Water heater, domestic, circulating | | | |
| or side-arm......................... | 35,000 | | |
| Refrigerator ......................... | 3,000 | | |
| Clothes dryer, domestic................ | 35,000 | | |

Btu/hr, and the heating value per cubic foot of gas is 1000 Btu. Assume the distance between the gas meter and the burner to be 75′.

By a substitution of values in the foregoing formula, we obtain:

$$\text{cu. ft. of gas/hr.} = \frac{155,000}{1000} = 155$$

From Table 9-2 it will be noted that a 1″ pipe 75′ long will handle 155 cu. ft./hr. with 0.3″ pressure drop. The determination of gas-pipe sizes for other heating appliance units can be made in a similar manner.

*Example:* It is required to determine the pipe size of each section and outlet of the piping system, with a designated pressure drop of 0.3″ water column. The gas to be used has a specific gravity of 0.60 and a heating value of 1000 Btu per cubic foot.

Reference to the piping layout and a substitution of values for maximum gas demand for the various gas outlets (Fig. 9-1) will be as follows:

$$\text{Gas demand for outlet A} = \frac{55,000}{1000} = 55 \text{ cu. ft./hr.}$$

$$\text{Gas demand for outlet B} = \frac{3,000}{1000} = 3 \text{ cu. ft./hr.}$$

$$\text{Gas demand for outlet C} = \frac{65,000}{1000} = 65 \text{ cu. ft./hr.}$$

$$\text{Gas demand for outlet D} = \frac{116,000}{1000} = 116 \text{ cu. ft./hr.}$$

**Table 9-2. Capacity of Pipe of Different Diameters and Lengths in Cu. Ft. per Hour with Pressure Drop of 9.3" and Specific Gravity of 0.60**

| Length of Pipe (ft.) | Iron-pipe sizes (IPS)(inch) | | | | |
|---|---|---|---|---|---|
| | $^1/_2$ | $^3/_4$ | 1 | $1^1/_4$ | $1^1/_2$ |
| 15 | 76 | 172 | 345 | 750 | 1220 |
| 30 | 52 | 120 | 241 | 535 | 850 |
| 45 | 43 | 99 | 199 | 435 | 700 |
| 60 | 38 | 86 | 173 | 380 | 610 |
| 75 | | 77 | 155 | 345 | 545 |
| 90 | | 70 | 141 | 310 | 490 |
| 105 | | 65 | 131 | 285 | 450 |
| 120 | | | 120 | 270 | 420 |
| 150 | | | 109 | 242 | 380 |
| 180 | | | 100 | 225 | 350 |
| 210 | | | 92 | 205 | 320 |
| 240 | | | | 190 | 300 |
| 270 | | | | 178 | 285 |
| 300 | | | | 170 | 270 |
| 450 | | | | 140 | 226 |
| 600 | | | | 119 | 192 |

From the piping layout in Fig. 9-1 it will be noted that the length of gas pipe from the meter to the most remote outlet (A) is 105'. This is the only distance used. Reference to Table 9-2 indicates:

Outlet A supplying 55 cu. ft./hr. requires ½" pipe;
Outlet B supplying 3 cu. ft./hr. requires ½" pipe;
Outlet C supplying 65 cu. ft./hr. requires ¾ pipe;
Outlet D supplying 116 cu. ft./hr. requires ¾ pipe;
Section 3 supplying outlets A and B, or 58 cu. ft./hr., requires ½" pipe;
Section 2 supplying outlets A, B, and C, or 123 cu. ft./hr., requires ¾ " pipe;
Section 1 supplying outlets A, B, C, and D, or 239 cu. ft./hr., requires 1" pipe.

305

**Fig. 9-1. Diagram of a gas-piping layout for a typical domestic installation.**

The determination of sizes of gas pipe for other piping layouts may be made is a similar manner.

## Diversity Factor

The diversity factor, or ratio of the maximum probable demand to the maximum possible demand, is an important factor in determining the correct size of gas piping to be used in multiple-family dwellings. It is dependent on the number and kinds of gas appliances being installed. Consult the servicing gas supplier or authority having jurisdiction for the diversity factor to be used.

## Piping Material

Standard-weight wrought-steel and wrought-iron pipe should be used in installation of appliances supplied with utility gases. Threaded copper or brass pipe in iron-pipe sizes may be used with gases not corrosive to such material. Gas pipe should not be bent. Fittings should be used when making turns in gas pipe.

The connection of steel or wrought-iron pipe by welding is

usually permissible. Threaded pipe fittings (except stop-cocks or valves) should be of malleable iron or steel when used with steel or wrought-iron pipe, and should be copper or brass when used with copper or brass pipe. When approved by the authority having jurisdiction, special fittings may be used to connect steel or wrought-iron pipe.

## Piping Material for Liquefied Petroleum (LP) Gases

Gas piping for use with undiluted liquefied petroleum (LP) gases should be of steel or wrought-iron pipe complying with the American Standard for wrought-steel and wrought-iron pipe, and brass or copper pipe, or seamless copper, brass, steel, or aluminum tubing. All pipe or tubing should be suitable for a working pressure of not less than 125 psi. Copper tubing may be of the standard K or L grade, or equivalent, having a minimum wall thickness of 0.032". Aluminum tubing should not be used in exterior locations or where it is in contact with masonry or plaster walls, or with insulation.

### Defective Material

Gas pipe or tubing and fittings should be clear and free from burrs and defects in structure or threading and should be thoroughly brushed, chipped, and scale blown. Defects in pipe, tubing, or fittings should not be repaired. When defective pipe, tubing, or fittings are located in a system, the defective material should be replaced.

Gas pipe, tubing, fittings, and valves removed from any existing installation should not be used again until they have been thoroughly cleaned, inspected, and judged to be equivalent to new material. Joint compounds should be applied sparingly and only to the male threads of pipe joints. Such compounds should be resistant to the action of liquefied petroleum (LP) gases.

### Pipe Threads

Pipe and fittings should comply with the American Standard for pipe threads. Pipe with damaged threads should not be used. All

307

gas piping should be threaded in accordance with specifications in Table 9-3.

## Branch Connections

All branches should be taken form the top or side of horizontal pipes and not from the bottom. When ceiling outlets are taken from horizontal piping, the branch should be taken from the side and carried in a horizontal direction for a distance of not less than 6″. Fig. 9-2 shows the right and the wrong way to connect a drop branch to a horizontal run.

### Table 9-3. Approximate Length of Thread Required for Various Sizes of Gas Pipe

| Size of pipe in inches | Approximate length of threaded portion, in inches | Approximate number of threads to be cut |
|---|---|---|
| $\frac{3}{8}$ | $\frac{5}{16}$ | 10 |
| $\frac{1}{2}$ | $\frac{3}{4}$ | 10 |
| $\frac{3}{4}$ | $\frac{3}{4}$ | 10 |
| 1 | $\frac{7}{8}$ | 10 |
| $1\frac{1}{4}$ | 1 | 11 |
| $1\frac{1}{2}$ | 1 | 11 |
| 2 | 1 | 11 |
| $2\frac{1}{2}$ | $1\frac{1}{2}$ | 12 |
| 3 | $1\frac{1}{2}$ | 12 |
| 4 | $1\frac{3}{4}$ | 13 |

WRONG WAY                    RIGHT WAY

NOT LESS THAN 6 IN.

Fig. 9-2. Connection of a drop branch to a horizontal pipe.

## Pipe Supports

Care must be taken to properly support the pipe and not subject it to any unnecessary strain. Fig. 9-3 shows a pipe supported by pipe clips or straps fastened to the building joists. The maximum distance these supports should be spaced depends on the pipe size. The following distances between supports should never be exceeded, with closer spacing being preferable.

$\frac{3}{8}$ " and $\frac{1}{2}$ " pipe ................................. 6'
$\frac{3}{4}$ " and 1" pipe .................................. 8'
1 $\frac{1}{4}$ " and larger (horizontal) ..................... 10'
1 $\frac{1}{4}$ " and larger (vertical) ........... every floor level

Pipe shorter than the support spacing listed should still be adequately supported. Wherever a branch fitting is used, or wherever there is a change of direction of 45° or more, a support should be provided within 6" of at least one side of the fitting or bend.

GAS PIPE     ATTACHED TO JOISTS
             WITH PIPE CLIPS

**Fig. 9-3. Gas piping should be supported from the joists and not on the wall.**

Pipe straps or iron hooks should not be used for fastening pipe larger than 2". Beyond this size, when the pipe is horizontal and is to be fastened to floor joists or beams, pipe hangers should be used; when the pipe is horizontal and is to be fastened to a wall, hook plates should be used.

When pipes run crosswise to beams, do not cut the beams to a depth of more than $\frac{1}{5}$ of the depth of the timber (Fig. 9-4). This cutting should be as near the support of the beam as possible (Fig. 9-5), but in no case should it be farther from a support than $\frac{1}{6}$ of the span. When possible, pipes should be run parallel to the beams

309

**Fig. 9-4.** The notches in joists should never be deeper than one-fifth the depth of the joist.

**Fig. 9-5.** Joists should be notched near where they are supported when it is necessary to run pipe across them.

to avoid cutting and weakening of the beams. Horizontal lines should have some pitch, as shown in Fig. 9-6, to provide for drainage of any condensed liquid, especially where pipes are exposed to cold. A T with a capped suspended nipple should be

310

provided at the lowest point to allow removal of the liquid. The size and length of the nipple should be in proportion to the amount of piping which drains into it. A cross instead of a T can be used to advantage in some installations. The use of a cross permits cleaning the horizontal and vertical runs, should the need ever arise. Fig. 9-7 shows a typical installation using a cross.

Fig. 9-6. Horizontal runs of gas piping should be slanted slightly, with a drip nipple installed on the lowest point to collect any condensate moisture.

Fig. 9-7. The use of a cross at the junction of a horizontal and vertical run permits cleaning either pipe as well as the installation of a drip nipple.

311

## Offsets

Where an offset in the line is necessary, such as a projection in a wall, the offset should be made with 45° elbows instead of 90° fittings. This procedure reduces the friction to the flow of gas, and also reduces the likelihood of stoppage in the line. Correct and incorrect offset installations are illustrated in Fig. 9-8.

Fig. 9-8. Offsets should be installed with 45° elbows instead of 90° fittings.

## Concealed Piping

Gas piping smaller in size than standard ½″ pipe should not be used in any concealed location. Concealed piping should be located in hollow portions rather than in solid portions. Tubing used with undiluted liquefied petroleum (LP) gases should not be run inside any walls or partitions unless they are protected against physical damage. This rule does not apply to tubing that passes through walls or partitions.

Gas piping in concrete floors should be laid in channels in the floor suitably covered to permit access to the piping with a minimum of damage to the building. When installing gas piping that is to be concealed, unions, tubing fittings, running threads, right- and left-hand couplings, bushings, and swing joints made by combinations of fittings should not be used.

When necessary to inset fittings in gas pipe that has been installed in a concealed location, the pipe may be reconnected by use of a ground-joint union with the nut center punched to prevent loosening by vibration. Reconnection of tubing to be installed in a concealed location is prohibited.

Gas piping should be protected against freezing temperatures. When piping must be exposed to wide ranges in temperature, special care should be taken to prevent stoppages.

## Closing of Outlets

Each outlet, including a valve or cock outlet, should be securely closed and made gas tight with a threaded plug or cap immediately after installation; it should be left closed until an appliance is connected. Similarly, when an appliance is disconnected from an outlet, the outlet should be securely closed and made gas tight to prevent leaks.

Outlets shall be far enough from floors, walls, and ceilings to permit the use of proper wrenches without straining, bending, or damaging the piping. Outlets should not be placed behind doors.

## Electrical Grounding

A gas piping system within a building should be electrically continuous and bonded to any grounding electrode, as defined by the National Electrical Code. The point of attachment of a grounding conductor to gas piping should always be on the street side of the gas meter and should be accessible where practicable.

## Gas Shut-off Valves

Manual shut-off valves controlling one or several gas piping systems should, in multiple systems, be placed at an adequate distance from each other so that they will be easily accessible for operation and should be installed so as to be protected from physical damage. Each valve should be plainly marked with a metal tag attached by the installing agency so that the gas piping systems supplied through them can be readily identified. It is advisable to place shut-off valves at every point where safety, convenience of operation, and maintenance demand them.

## Gas Regulator and Gas Regulator-Vent Outlets

Every gas service connection supplying gas into any building at a pressure in excess of one pound per square inch should be provided with a regulator that will reduce the pressure of such gas prior to its entering the meter to not more than ½ pound per square inch; higher pressure may be allowed for commercial or industrial use.

Each gas regulator should be provided with a ventilating pipe that should lead directly to the outer air, and the outlet, where practicable, should not be located under a window or any opening leading back into the premises. Covering over, plugging up, or otherwise obstructing any gas regulator-vent outlet should be unlawful. A gas vent identified by suitable marking should be attached to the outlet on the outside of the building.

## Gas Service Connection

Every gas service connection brought into a structure should be fitted with an approved lubricated, straight-way shut-off stop-cock or shut-off valve, or equivalent, constructed so that the core cannot be blown out by the pressure of the gas. The stop-cock or shut-off valve should be placed in an accessible position immediately inside the street side of the gas meter and of the gas regulator, if any.

In all high-pressure areas, the gas company concerned should at least once each year inspect the stop-cock or shut-off valves to ensure that they are all in good working order and ready for immediate use. All materials used in the installation of stop-cock or shut-off valves should be approved by local authorities. All matters in relation to stop-cock or shut-off valves not covered by this section should be determined by the authorities.

## Test Methods

Before appliances are connected, piping systems should stand a pressure of at least 6″ of mercury or 3 lbs. of gauge pressure for a period of not less than 10 minutes without showing any drop in pressure. Pressure should be measured with a water manometer or an equivalent device calibrated to read in increments of not greater than ⅒″ water column for a period of not less than 10 minutes without showing any pressure drop. The source of the pressure should be isolated before the tests are made.

## Purging

After gas piping has been checked, the piping receiving the gas should be fully purged. One method of purging gas-appliance piping is to disconnect the pilot piping at the outlet of the pilot valve. Under no circumstances should piping be purged into the combustion chamber of an appliance.

# LIQUID PETROLEUM GAS

Liquefied petroleum gas is a fuel made up principally of propane or butane, or a mixture of these products. At atmospheric pressure pure butane returns to gas, or vaporizes, at a temperature of 32°F. Pure propane vaporizes or returns to gas at 44°F. Butane is more widely used in mild climates, propane can be used in much colder climates. LP gas is used in many areas not served by natural gas mains. It is also used in many portable applications such as campers and service trucks. LP gas has some advantage over natural or manufactured gas and also some disadvantages. The principal advantage is the higher Btu content of LP gas. The heat content of LP gas in roughly three times greater than natural gas and six times greater than manufactured gas. The principal disadvantage is that LP gas is heavier than air and therefore extreme care must be taken when installing tanks and piping for LP gas.

Natural gas, which is lighter than air, will escape upward in the event of leakage. LP gas, which is heavier than air, will drop from the point of leakage and may collect in a low place or pocket and remain there for a considerable length of time. If a spark or flame should come into contact with the gas, an explosion could occur. More than a few LP gas explosions have been traced to a leakage of LP gas.

LP gas tanks and cylinders must be installed or stored outside any building used for people, in a well-ventilated area and placed so that they are readily accessible at all times for inspection, testing and shutting off the gas supply. All service-piping and main-supply shut-off valves should be outside the building and readily accessible. Special pipe-joint compounds that will not deteriorate or dissolve when in contact with LP gas should be used on all threaded connections. All LP gas installations must comply with local codes or regulations.

## USING A MANOMETER

Although manometers can be used for several purposes, the plumber and pipe fitter will find the manometer useful for testing gas piping and for measuring the gas pressure delivered to the appliances in a building. In almost all areas using natural gas, the gas is delivered to a regulator located on the main side of the meter at a higher pressure than the appliances in the building are designed for. The higher pressure is reduced at the regulator and then passed on through the meter to the appliances. The pressure may vary somewhat from one area to another; in general use, the pressure on the downstream side of the regulator will average from 5″ to 6″ of water column. In order to check the gas pressure it will be necessary to connect the hose from one side of the manometer to the gas piping. This can be done at a gas stove by removing a burner and sliding the manometer hose on the burner connection or by connecting the manometer hose to the piping at any other desirable point. The manometer is prepared for use by first pouring water into the tube on one side until it equalizes at the zero point on the gauge. To ensure an accurate reading this should be done carefully; it may be necessary to add water two or three times in order to equalize the water in the tube. When the water level is equalized at the zero point, connect the manometer to the test point. Turn the gas on at the test point and observe the reading on each side of the manometer tube. Fig. 9-9A shows a manometer with the water equalized at the zero point. Fig. 9-9B shows the water column after the gas has been turned on at the test point. The reading at the left side of the tube is $2\frac{1}{2}$ below the zero point. The reading on the right side of the tube shows $2\frac{1}{2}$ above the zero point. The two readings are added together, $2\frac{1}{2} + 2\frac{1}{2} = 5$. The pressure at the test point is 5″ of water column pressure. The readings of both sides of the manometer tube are always added together to obtain the pressure $(2 + 2 = 4)$ $(3\frac{1}{2} + 3\frac{1}{2} = 7)$. When a manometer is used for testing pressure, one side of the tube is always left open, the other side is connected to a convenient

THIS SIDE OF TUBE OPEN

TUBE CONNECTED TO
TEST POINT

A

B

**Fig. 9-9. Using a manometer.**

point to make the test. Gas pressures may vary somewhat due to the usage at a particular time or season, but if the manometer shows a greater deviation from normal than should be expected, the utility or supplier should be notified so that corrective action may be taken.

317

One of the causes of low gas pressure is a stopped-up vent opening from a regulator. A regulator, to work properly, must "breathe"; the diaphragm flexes or expands outward each time gas flows through the meter, and contracts inward when the gas flow ceases. If the vent opening becomes plugged by snow, ice, or any other cause, the diaphragm will not be able to flex, or open, and little or no gas will be able to enter the piping. In the event of low gas pressure, the first item to check is the regulator.

CHAPTER 10

# Air Conditioning

Year-round air conditioning is designed to control the quality of indoor air during the entire year. Summer air conditioning lowers the air temperature, filters, cleans the air, and dehumidifies it. Winter air conditioning heats the air, cleans and filters it, and adds humidity. (Most people, of course, only think of *cooling* as air conditioning.)

Three factors are important in summer air conditioning: temperature, relative humidity, and air circulation. Relative humidity is the percentage of moisture in the air, compared to the greatest amount (100%) the air can hold. The warmer the air, the more moisture it can hold.

If the temperature of the air in a building is 80°, the air is *capable* of holding 12 grains of moisture per cubic foot. If the moisture content of the air is measured and found to only 6 grains percubic foot, the relative humidity is 50%. If the same air is passed through a cooling coil and cooled to a temperature of 60° with 6 grains per cubic foot, the relative humidity of the air is 100% since air at 60° is capable of holding only 6 grains per cubic foot.

When air is cooled to the point where its relative humidity is 100%, it is at the dew point.

If the 60° air is cooled to 50° as it passes through the cooling coil, it can hold only 4 grains of moisture per cubic foot. Two grains of moisture will be condensed into water as it passes over and through the cooling coil. In a typical air-conditioned building in the summer, cooled, dried, and cleaned air is circulated, through the supply ducts, into the room spaces. As the cool dry air circulates it absorbs heat and moisture and is returned through the return ducts to the filters and then to the cooling coil, or evaporator. When the warmed moist air passes over and through the cooling coil, it is cooled, and the excess moisture in the air is condensed into water, which drips off the coil and is drained away.

All refrigeration systems, whether operated by gas or electricity, depend on two basic principles:

1. *A liquid absorbs heat when it boils or evaporates to a gas.* Examples:

   a. Freon-12 boils to vapor at −21° at atmospheric pressure.
   b. Water boils to a gas (steam) at 212° at atmospheric pressure.
   c. Liquid ammonia boils to vapor at −28°F at atmospheric pressure.

2. *As vapor or gas condenses to a liquid form, heat is released.* One of the important rules of refrigeration is: The boiling point of a liquid is changed by changing the pressure the liquid is under. *Raising* the pressure *raises* the *boiling point. Lowering* the pressure *lowers* the *boiling point.*

## COOLING TOWERS

Water is used as a cooling medium in some air-conditioning systems. It is most practical and economical to reuse this water, and since the water absorbs heat in the condenser, it must be cooled in order to reuse it. Fig. 10-1 shows a schematic diagram of a water

**Fig. 10-1. Use of a cooling tower in an air-conditioning system that uses a water-cooled condenser.**

air-conditioning system using a cooling tower. The cycle of operation is basically simple: cooled water from the cooling tower basin is pumped through the water jacket of the condenser. The cooled water absorbs heat from the hot gas in the refrigerant piping coils in the condenser. The water, now hot from the absorbed heat, is circulated to the cooling tower where it runs into a distribution trough and drips down through the baffles of the tower. A fan pulls air into the tower, circulating the air into, through, and around the baffles. The water dripping through the baffles gives off heat into the air, and the heated air is pushed out of the exhaust opening in the tower. The cooled water drips into the basin in the base of the tower, ready to be circulated back to the condenser to again absorb heat. A small amount of water is lost by evaporation in the tower, and a make-up water valve is installed in the tower basin to replace the lost water. Fig. 10-2 shows a cutaway drawing of a cooling tower, and Fig. 10-3 shows a typical cooling-tower installation.

## ELECTRICAL AIR CONDITIONING

The operation of an electrical (compressor) air-conditioning system using an air-cooled condenser is essentially the same as one using a water-cooled condenser, the only difference being that air, pulled through the condenser by a fan, is used as the condenser's

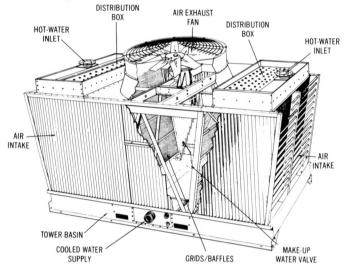

Fig. 10-2. Cutaway view of a cooling tower.

Fig. 10-3. Cooling towers used with water-cooled condensers.

cooling medium. The path of the refrigerant and the changes that take place in the refrigerant as it passes through the compressor, condenser, expansion valve, and evaporator are explained elsewhere in this chapter. Fig. 10-4 shows a schematic of an electrical air-conditioning system.

**Fig. 10-4. Electrical air-conditioning system.**

A typical electric air conditioner operates by passing a cold liquid-plus-vapor mixture, at a lowered pressure, through an evaporator, or cooling coil. The cold liquid-plus-vapor mixture absorbs heat from the air passing over the cooling coil. As the liquid-vapor passes through the coil, absorbing heat, it is completely vaporized, changing from a liquid to a cold gas. The compressor pulls the cold gas from the evaporator or cooling coil as fast as it vaporizes, through the *suction line.* Next, the compressor compresses the cold gas to a higher pressure and in the process also raises the temperature of the cold gas, changing it to a hot gas. The increased pressure forces the hot gas through the condenser where either air or water removes the heat from the refrigerant gas. The hot gas must be at a higher temperature than the air or water in order for heat to flow from the gas to the air or water. As the hot gas gives off heat in the condenser it *condenses,* becoming a hot *liquid,* and still under high pressure. The liquid refrigerent flows from the condenser to an expansion valve or a capillary tube, and then passes through the valve to the evaporator. The pressure up to the expansion valve is higher than the pressure in the evaporator; therefore, as the liquid refrigerant passes through the expansion valve the refrigerant expands rapidly and partially vaporizes as it enters the evaporator. Because the pressure is reduced at the evaporator, the temperature is also reduced, changing the refrigerant from a hot liquid to a cold liquid-plus-vapor mixture. The cold liquid-plus-vapor mixture can now absorb the heat from the air passing over and through the evaporator (cooling coil) and the process of air conditioning continues. The diagram of an electric air-conditioning unit (Fig. 10-4) explains the steps in the cycle of operation and the changes that take place in the refrigerant as it goes through the cycle.

It is important to remember that an air-cooled condenser operates because the refrigerant is at a higher temperature than the air passing through the condenser and thus the heat absorbed by the refrigerant in the evaporator can flow to the cooler air at the condenser.

## UNIT AND CENTRAL
## AIR-CONDITIONING SYSTEMS

The installation, maintenance, and repair of air-conditioning systems has become an important and profitable addition to the heating and plumbing trade. The demand for central, full-house air conditioning installed as an integral part of the heating system has been largely responsible for the acceptance of air conditioning as a natural part of the heating and plumbing trade. The installation of commercial air conditioning also involves the use of many plumbing techniques, again emphasizing how closely air conditioning is allied with plumbing and pipe fitting.

Air conditioning may be classified in two general groups—*unit* systems and *central* systems. The unit air conditioner is placed in the room to be cooled and, as a rule, contains the necessary fans, filters, compressor, and controls, all contained within a specially designed case. It requires a minimum of installation with no plumbing involved, with the exception of a possible drain to remove any condensed moisture from the unit. Window air conditioners (Fig. 10-5) are the most familiar of the unit air conditioners, but are normally placed in the appliance category.

Central air-conditioning systems are located at some central point in the building to be cooled, and distribute the cooled air to the various parts of the building through a duct system. The system may be completely separate from the heating system, especially in commercial buildings. The majority of central air-conditioning systems designed for homes, however, utilize the existing duct work of warm-air furnaces to cool the home. Where steam, hot-water, or radiant-heating systems are used, a separate cooling system is necessary and will require the installation of ducts to carry the cool air to the various rooms.

Combination heating and cooling systems are popular and offer the most economical means of year-round comfort in the home. Many of these units automatically switch from heating to cooling as required without any attention from the occupants other than setting a thermostat to the temperature desired. Some units require a switch to be thrown to select either heating or cooling.

Courtesy Gibson Refrigerator Sales Corporation

Fig. 10-5. A typical room air conditioner.

325

# CAPACITY REQUIREMENTS

The important variables to keep in mind when estimating the Btu requirements for an air-conditioning installation are:

1. The total square feet of floor area.
2. The type of wall and ceiling construction.
3. The proportion of the outside wall area that is glass.
4. The walls of the space to be air conditioned that are exposed to the sun.

Additional factors to be taken into account are ceiling heights, number of persons using the space, difference in temperature between inside and outside, and miscellaneous heat loads such as lamps and radio and television sets.

The cooling capacity of an air-conditioning system is its ability to remove heat from the building, and is usually measured in Btu's per hour for the smaller units. The higher the Btu rating, the more heat can be removed. The capacity rating of a unit is normally given on the nameplate, together with such other necessary information as voltage and wattage requirements. It is important that the unit be large enough to adequately cool the required space without running at full capacity for extended periods.

When dealing with large air conditioners, especially those used for central cooling and for commercial installations, the term *ton of refrigeration* is most often used instead of a Btu rating. A ton of refrigeration is equivalent to 12,000 Btu/hr. This is derived from the fact that one pound of ice absorbs 144 Btu when it melts. Thus, one ton of ice absorbs 2000 × 144, or 288,000 Btu. When one ton of ice melts in 24 hours, the rate is 288,000 ÷ 24, or 12,000 Btu/hr.

In residences and offices, a one-ton unit will usually take care of 5000 to 7000 cubic feet of space, depending on the number of occupants. A one-ton unit in a theater will take care of approximately 15 seats. Thus, for a theater seating 1500 people, a 100-ton machine would be needed. The volume of such a theater is around 800,000 cubic feet, meaning that one ton of refrigeration is required for each 8000 cubic feet in an average theater.

## Energy Efficiency

Since energy consciousness has taken hold in the country, a wide variety of energy changes have been made. One of the most effective is the practice of rating various appliances for energy efficiency. By government mandate, air conditioners, central air-conditioning systems, and other units must be accompanied by a fact sheet that contains the unit's energy efficiency ratio, or EER, and how much it costs to operate the unit. It becomes a simple matter to compare and contrast the efficiency of various units, as Fig. 10-6 shows.

## Ventilation

Practically all rooms and buildings have a certain amount of natural ventilation, called *infiltration*. Air seeps through cracks in the floors and walls, and around windows and doors, amounting to a surprising number of cubic feet of air per hour. The most practical method of estimating infiltration is to base it on the cubical content of the space to be cooled. For the average home, office, and small shop where only a limited number of people gather, natural ventilation is usually sufficient. Table 10-1 lists the natural changes that can be expected for normal construction.

A fault commonly found with air conditioning is drafts. Where comfort is concerned, the conditioned air must be evenly and thoroughly distributed without a trace of draft.

## Estimating Infiltration

If it is desirable to estimate the sensible heat load of a space, room, or building, use the following formula:

$$H_s = \frac{V(t_o - t)}{n \times 3360}$$

where,

$H_s$ is the sensible heat in Btu per minute from infiltration,

V  is the volume of the space, room, or building in cubic feet,

$t_o$ is the outdoor dry-bulb temperature assumption,

t  is the indoor dry-bulb temperature to which the space is to be cooled,

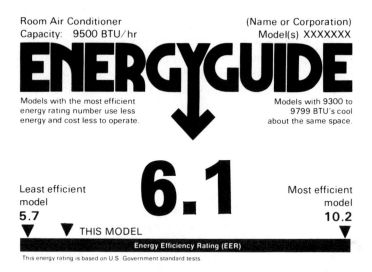

Room Air Conditioner
Capacity: 9500 BTU/hr

(Name or Corporation)
Model(s) XXXXXXX

# ENERGYGUIDE

Models with the most efficient
energy rating number use less
energy and cost less to operate.

Models with 9300 to
9799 BTU's cool
about the same space.

Least efficient
model
**5.7**
▼          ▼ THIS MODEL

# 6.1

Most efficient
model
**10.2**
▼

**Energy Efficiency Rating (EER)**

This energy rating is based on U.S. Government standard tests.

## How much will this model cost you to run yearly?

| Yearly hours of use | | 250 | 750 | 1000 | 2000 | 3000 |
|---|---|---|---|---|---|---|
| | | Estimated yearly $ cost shown below | | | | |
| Cost per kilowatt hour | 2¢ | $ 8 | $ 23 | $ 31 | $ 62 | $ 94 |
| | 4¢ | $ 16 | $ 47 | $ 62 | $ 125 | $ 187 |
| | 6¢ | $ 23 | $ 70 | $ 94 | $ 187 | $ 281 |
| | 8¢ | $ 31 | $ 94 | $ 125 | $ 250 | $ 374 |
| | 10¢ | $ 39 | $ 117 | $ 156 | $ 312 | $ 468 |
| | 12¢ | $ 47 | $ 140 | $ 187 | $ 374 | $ 562 |

Ask your salesperson or local utility for the energy rate (cost per kilowatt hour) in your area. Your cost will vary depending on your local energy rate and how you use the product.

Important Removal of this label before consumer purchase is a violation of federal law (42 U.S.C. 6302)

(Part No. 20648)

**Fig. 10-6. Energy label required on air-conditioning units.**

n   is the number of hours required to effect one complete air change.

The total heat load of infiltration can be computed by means of the following formula:

$$H_t = \frac{V(H_o - H_1)}{n \times 810}$$

$H_t$ is the total heat load in Btu per minute due to infiltration,
$H_o$ is the total heat per punds of air at the outdoor wet-bulb temperature,
$H_t$ is the total heat per pound of air at the indoor wet-bulb temperature.

If it is desired to estimate the latent heat load, use the following formula:

$$H_1 = H_t - H_s$$

where,
$H_1$ is the latent heat load,
$H_t$ is the total heat load,
$H_s$ is the total sensible heat load.

### Refrigeration Demand

Table 10-2 may be used to calculate the amount of refrigeration required to cool the air and reduce the humidity. Note that the table lists the load demand in Btu per cubic foot of air to be cooled a certain number of degrees and removing moisture to a specified humidity level.

Table 10-1 lists the number of air changes dependent on the exposure. For excellent construction, the minimum infiltration figure can be employed, and for old or poor construction, the maximum factor should be used. This table can be used in estimating the cooling and condensation load by multiplying the number of air changes by the volume of the room, and then multiplying this result by the proper factor from Table 10-2. This will give the total Btu cooling load required, and provides a very convenient and reasonably accurate method.

For greater accuracy in estimating the cooling load, it is suggested that the installer consult appropriate manuals published by

329

## Table 10-1. Infiltration.

| Exposure | Changes per Hour |
|----------|------------------|
| One Side | $\frac{1}{2}$ to 1 |
| Two Sides | $\frac{3}{4}$ to $1\frac{1}{2}$ |
| Three Sides | 1 to 2 |
| Four Sides | 1 to 2 |
| Inside Room | $\frac{1}{2}$ to $\frac{3}{4}$ |

## Table 10-2. Refrigeration Load for Air Cooling

| Relative Humidity | Temperature Reduction Increments, °F | | | | | | | | | | |
|---|---|---|---|---|---|---|---|---|---|---|---|
| % | 5° | 6° | 7° | 8° | 9° | 10° | 11° | 12° | 13° | 14° | 15° |
| 40 | 0.045 | 0.085 | 0.120 | 0.155 | 0.195 | 0.235 | 0.280 | 0.320 | 0.365 | 0.415 | 0.460 |
| 45 | 0.123 | 0.165 | 0.205 | 0.245 | 0.290 | 0.335 | 0.383 | 0.428 | 0.482 | 0.530 | 0.580 |
| 50 | 0.200 | 0.245 | 0.290 | 0.335 | 0.385 | 0.435 | 0.485 | 0.535 | 0.590 | 0.645 | 0.700 |
| 55 | 0.280 | 0.352 | 0.377 | 0.425 | 0.480 | 0.528 | 0.585 | 0.642 | 0.703 | 0.765 | 0.825 |
| 60 | 0.360 | 0.410 | 0.465 | 0.515 | 0.575 | 0.630 | 0.685 | 0.750 | 0.815 | 0.885 | 0.950 |
| 65 | 0.438 | 0.488 | 0.543 | 0.598 | 0.663 | 0.720 | 0.783 | 0.850 | 0.918 | 0.990 | 1.170 |
| 70 | 0.515 | 0.565 | 0.620 | 0.680 | 0.745 | 0.810 | 0.875 | 0.950 | 1.020 | 1.095 | 1.175 |
| 75 | 0.587 | 0.648 | 0.703 | 0.763 | 0.833 | 0.895 | 0.968 | 1.055 | 1.125 | 1.202 | 1.295 |

the Air Conditioning and Heating Contracters Association in Washington D.C. For residential use, manual J is used; commercial or large units are covered in manual N. These manuals take in a variety of factors, including the construction of the walls, heat loss, solar radiation, and items within a structure, such as motors, that generate heat.

## MOISTURE

Moist or properly humidified air serves to maintain paper, wood, plaster and plasterboard, rugs, and cloth at the correct moisture balance so that such materials are kept in the proper condition. Dehydration increases the fragility of most materials and, in many cases, causes a change in texture and color. Under ordinary conditions, when the outdoor temperature makes it necessary to use the heating system, materials in the heated space will give up their moisture to the air. This fact can be shown by the condensation on windows during frosty days of the heating season. During this time, rugs, plaster, wood, and the like give up a certain portion of their moisture content.

Humidification (addition of moisture to the air) is usually necessary during such periods. When the humidity is increased, the materials in the heated space will absorb the amount of moisture lost during the dehydrating period. The absorption, and especially the reabsorption, of moisture by these materials is known as the *moisture regain*.

Comfort is also affected by the amount of moisture in the air. During the heating season, if the humidity is allowed to decrease too much, the dry air becomes uncomfortable and feels chilly, even though its temperature may be normal. Thus, a humidity level of 30% to 50% must be maintained to ensure comfort. When the humidity is kept at this level, condensation or frosting occurs on the windows and door glass. This is objectionable for many reasons. If the condensation is appreciable and remains for extended periods, it will damage or destroy the finish on the window frames and sills and will create rust and rot. Putty or glazing compound is also impaired and may have to be replaced periodically.

Condensation can be prevented by maintaining a lower relative humidity, but this will result in discomfort as well as damage to other material in the heated space. Instead, condensation can be reduced or eliminated by the use of double or triple window glass; this also results in energy savings.

Double glass, with dead-air space between, will result in the inside glass having a higher surface temperature than where a single glass is employed, so that little or no condensation will occur under normal conditions. Table 10-3 lists the temperature of the glass at which condensation takes place for various relative humidities.

## COOLING EQUIPMENT

It is possible to select an air-conditioning unit or units to suit practically any building. The cooling apparatus may be entirely separate from the heating equipment, may be added to the equipment, or may be included as part of the heating plant. In the

### Table 10-3. Temperature and Relative Humidity at Which Condensation Occurs on Window Glass

| Outdoor Temperature | Temperature of Inner Surface | | Percent of Relative Humidity at Which Condensation Forms | |
|---|---|---|---|---|
| °F | Single glass | Double glass | Single glass | Double glass |
| −10 | 13 | 46 | 12 | 45 |
| 0 | 20 | 49 | 18 | 49 |
| 10 | 28 | 52 | 23 | 54 |
| 20 | 34 | 54 | 30 | 60 |
| 30 | 42 | 57 | 38 | 68 |
| 40 | 48 | 60 | 48 | 75 |
| 50 | 55 | 63 | 64 | 83 |
| 60 | 62 | 66 | 80 | 91 |
| 70 | 69 | 69 | 99 | 99 |

last case, a true air-conditioning system is the result, giving the occupants heat in winter and cooling in the summer without the necessity of switching from one to the other. It is all taken care of by automatic controls that require only the setting of a thermostat to the temperature desired.

Fig. 10-7 shows an air-conditioning cooling system for use in conjunction with a warm-air furnace. The unit shown contains the condenser, compressor, and fan to cool the condenser. This unit is normally installed in some remote location where the heat it must dissipate will not interfere with the space to be cooled. A unit of this type is often installed outside the building. The cooling coil is a separate unit and is installed in the furnace plenum. Flexible copper refrigerant lines connect the cooling coil and the condensing unit, being brazed in place. These lines are available in various lengths. Special couplings permit both the cooling coils and the condensing unit to be fully charged and hermetically sealed at the factory. The connections to the condenser unit are shown at the lower left corner of the cabinet in Fig. 10-7.

Many refrigerating units incorporate a sealed compressor similar to the one shown in Fig. 10-8. This type of compressor is quite popular for the smaller air-conditioning installations such as those for homes and small commercial buildings. The built-in motor contributes to the quiet operation of a compressor of this type, since a separate motor and drive belt are unnecessary. Spring suspension absorbs nearly all vibration. This particular model

Courtesy Janitrol Division, Midland-Ross Corp.

**Fig. 10-7. Condenser-compressor unit for use with an existing warm-air furnace.**

features a three-stage muffler and three expansion chambers to reduce noise level and discharge-gas pulsations.

Where it is impractical or undesirable to install cooling coils in the plenum of a warm-air furnace, individual cooling cabinets can be used in various locations in the space to be cooled. Fig.10-9 shows such a unit. These cabinet units provide custom-tailored conditioned air to meet nearly any requirement for individual rooms. Many sizes and styles are available for floor, wall, and ceiling installations. These units are also available with various types and sizes of heating coils to provide year-round air conditioning. Each unit can be individually controlled by selecting the proper control from the many that can be supplied. The unit in Fig. 10-9 has the back panel removed to show the cooling-coil assembly, the blower motor and fans, and the removable air filter.

Courtesy Trane Company

Fig. 10-8. A sealed compressor is used in many small air-conditioning systems.

Courtesy Dunham-Bush, Inc.

Fig. 10-9. A cabinet-type cooling unit with back cover removed.

For air conditioning commercial and industrial buildings, larger units are necessary. Roof-mounted systems are becoming quite popular, especially where inside space is at a premium. Combining both heating and cooling, this type of installation eliminates the need for a heating and cooling room. Fig. 10-10 shows the interior of a typical roof-mounted unit in which a gas burner is incorporated for heating purposes. Models are available without the heating section where summer cooling only is required. Electric heat is also available instead of gas. The unit is completely self-contained in a weatherproof and insulated cabinet. Systems from 15 to 35 tons of refrigeration are manufactured to fill a wide range of air-conditioning needs.

The units described previously have all utilized air as the cooling medium for the condenser. Water-cooled condensers are required

Fig. 10-10. Interior view of a roof-mounted air conditioner. This model features a gas-fired heating unit as well as a refrigeration unit.

or are desirable for some air-conditioning installations. The piping diagram of a typical roof-mounted water-cooled packaged air-conditioning unit is illustrated in Fig. 10-11.

A water-cooled condenser and compressor unit is shown in Fig. 10-12. This unit is for use when it is desirable to locate the cooling coils remotely from the compressor assembly.

For cooling very large buildings, a separate compressor room is often necessary. An example of such an installation is shown in Fig. 10-13. This system is used to cool a large shopping center in eastern New York State.

## CONTROL DEVICES

Three types of controls are used with air-conditioning installations. These are:

1. Manual.
2. Automatic.
3. Semiautomatic.

The manual control requires personal attention and regulation at frequent intervals, especially when the load varies continuously. Even if the system is in operation only during certain times of the day, inside and outside conditions usually vary so much that constant adjustments are necessary.

The manual control may consist of an ordinary on-off switch, which must be operated by hand to start and stop the actuating mechanism. This type of control is seldom used.

Automatic controls will meet the demands of a varying load almost instantly. An automatic system usually involves the use of several instruments, such as time clocks, thermostats, humidity controllers, and automatic dampers operated either by electricity or by compressed air. Automatic controls provide all the features considered necessary or desirable for temperature and humidity regulation as well as economy of operation.

A system using only a thermostat is usually a semiautomatic system and, in most cases, will provide satisfactory operation. Time clocks may be desirable on installations where definite shutdown periods are required. The fresh-air intake from outdoors should always be equipped with adjustable louvers so that the air

TEMPERATURE CONTROL  WATER OUT  SUPERHEATER  INSULATED SUCTION LINE  TX VALVE  COOLING COIL

DRAIN

WATER IN

ELECTRICAL CONTROLS  SIGHT GLASS  LINE VALVE  REFRIGERANT IN CONN.  VIBRATION ELIMINATOR

CONDENSER

REFRIGERANT DRIER-STRAINER  SOLENOID VALVE  REFRIGERANT OUT CONN.  SPRING MOUNT  COMPRESSOR

Courtesy Dunham-Bush, Inc.

**Fig. 10-11. A typical packaged air-conditioning unit.**

Fig. 10-12. A compact industrial cooling unit.

Courtesy Dunham-Bush, Inc.

Fig. 10-13. A large air-conditioning installation for a shopping center.

intake can be regulated for both summer cooling and winter heating. It is desirable to keep the control system on any installation as simple and free from service liabilities as possible.

## Thermostats

The device that operates by changes in room temperature is called a *thermostat*. Some thermostats depend on a bimetallic strip to make and break a set of contacts. In other designs, the bimetallic strip is coiled into a loose helix, with the end of the outer coil rigidly fixed to the case. The inner end of the coil is free to move according to temperature changes, and to this free end is attached a glass capsule containing mercury. The ends of two wires project into and are sealed in one end of the capsule so that, when it is tilted, the mercury flows to the low end. If the low end contains the projecting wire ends, the circuit is completed; if tilted the other way, the circuit is broken. The mercury bridges the gap between the two wires and acts as a metallic conductor.

The capsule is usually charged with some inert gas so that no arcing will occur when the electrical circuit makes and breaks. Such devices are made to handle currents up to and including that taken by a 1-HP motor. For larger motors, a relay used in conjunction with a thermostat is employed.

Another variety of thermostat depends on the pressure-temperature relationship of a gas for its operation. In this type, a bulb is charged with refrigerant and sealed. A small tube connects the charged bulb to an expandable diaphragm or drum. With an increase in temperature, there will be a corresponding increase in pressure. By means of levers, springs, and adjustments, the device can be set to cut in or out at any desired point.

A thermostat may be used to stop and start an air-conditioning system equipped with either an automatic constant-pressure expansion valve or a thermostatic expansion valve. If cooling only is desired, the simplest arrangement is to connect a single-pole, single-throw thermostat in series with the solenoid coil of a magnetic liquid-line stop valve in the liquid line of the evaporator expansion valve. Such a thermostat should be connected in series with a snap switch to make sure the thermostat does not operate the equipment during shutdown periods.

For circulating cold water or brine, a thermostat may be connected in series with the solenoid coil of the circulating-pump starting switch. With this arrangement, one thermostat can control the circulation of the cooling medium to a number of cooling cabinets.

## Humidity Controls

The instrument used to automatically control humidity is known as a *hygrostat* and is available under a variety of trade names. The element in the control is some hygroscopic material such as wood, human hair, or paper. When the material absorbs water from the air, it expands; when the material gives up moisture, it contracts. It is this action that is used to make and break an electrical circuit controlling the humidifying or dehumidifying equipment.

## Time Clocks

Time clocks are merely clock-operated switches placed in series with the thermostat to automatically prevent operation except during certain desired periods.

## Damper Motors

Motors for operating dampers are provided when automatic regulation of air flow is required in heating, ventilating, and air-conditioning systems and supply ducts. The motors are usually powered by electricity (compressed air is sometimes used) and controlled by thermostats. The dampers can be positioned by the motors as required and in accordance with the indication from the governing device.

Dampers can also be operated by thermostatic elements in which the expansion and contraction of a liquid or gas supplies the power.

## Contactors, Starters, and Relays

Most condensing units having motors above 1½ HP require a starter or contactor. This device will usually be furnished as part of the unit. Relays are necessary in many control and pilot circuits. These electrically operated switches are similar in design to contactors, but are generally lighter in construction and carry a smaller current.

## Limit Switches

Limit switches are often used to prevent the fan motor in a heating installation from circulating cold air when the heat is off, to prevent a humidifier from operating when the heat is off, or to hold refrigeration off when the heat is on.

## Heat Pumps

Back in the 1950's, the heat pump (Fig. 10-14) came onto the heating/cooling scene, but as time went by it faded, probably mainly because manufacturers were not careful about quality control and servicing. Since the early 1970's, however, and the

Courtesy General Electric

Fig. 10-14. Heat pump—it heats in winter, cools in summer.

energy crunch, the heat pump has made its way back on the scene and may be even more of a factor in the not-too-distant future.

It should be stated at the outset that deciding whether a heat pump is worth it in an individual installation is something that requires very careful evaluation and more than a dash of cynicism about claims one is bound to hear.

Basically, a heat pump pumps heat. In the winter, it draws heat from outside air (or water) and pumps it into the house—no fossil fuel is required. In the summer, it acts as an air conditioner, pumping the heat inside the house to the outside. In the typical installation, there is a compressor outside the house that is housed in a metal jacket, or box. A refrigerant (Freon) runs in pipes from the compressor to a heat exchanger inside the box in the home. When the refrigerant enters the compressor, it is chilled; but it absorbs heat from the atmosphere and becomes a gas. It is squeezed by a piston, which raises the temperature of the gas to around 100°. This gas enters the house where the heat exchanger "absorbs" the heat; this heat is then blown throughout the house through duct work.

In the summer, the reverse occurs, and the heat is drawn from the house and the house (or other installation) is cooled.

One problem with air-source heat pumps is that the colder the weather gets, the more difficulty the pump has in drawing heat from the atmosphere. Experts say that when the temperature gets much below 20°, the heat pump stops working and an electric resistance heater inside it switches on. The overall efficiency, or cost saving, becomes moot.

As might be expected, heat pumps sell better in the South, where temperatures do not go down as much as in the northern states.

The water-source heat pump draws heat from water that is 45° or above—from ground water. As it happens, even water in the coldest climates can be that warm as long as it is below the frost line (Figs. 10-15 and 10-16).

Companies make large claims for heating/cooling efficiency of a heat pump with a water source, but it does bear checking out, and factored into it must be the problem of the water. You not only need a source of from 5 to 15 gals. a minute, but a place to reroute it after the warmth is drawn from it. Some people use water from a

Courtesy Dunham-Bush, Inc.

**Fig. 10-15. How a heat pump cools.**

Courtesy Dunham-Bush, Inc.

**Fig. 10-16. How a heat pump heats.**

well and reroute it there, while others pump the water into a drywell.

You may also need duct work—larger than standard sizes—though the ducts in the average home are often oversized to begin with.

# SERVICING AIR CONDITIONERS

A malfunctioning system may be caused by one part of the system or a combination of several parts. For this reason, it is necessary and advisable to check the more obvious causes first. Each part of the system has a definite function to perform, and if this part does not operate properly, the performance of the entire air-conditioning system will be affected.

To simplify servicing, especially on a system that is unfamiliar, the serviceman should remember that the refrigerant, under proper operating conditions, travels through the system in one specified direction. Remembering this, the path of the refrigerant can be traced through any refrigerating system. For instance, beginning at the receiver, the refrigerant will pass through the liquid shut-off valve. If this valve is partially closed, or is completely closed or clogged, no refrigeration is possible, even though the unit itself can and will operate.

## Strainers and Filters

Most refrigerating units have a strainer or filter in the liquid line. If this device becomes clogged or filled with dirt, the liquid refrigerant will be unable to pass. Filters are usually designed to hold a certain amount of dirt, scale, metallic particles, and the like, and still function; but sometimes, through carelessness, an excessive amount of extraneous matter may be allowed to enter the system during assembly or servicing. The remedy is obvious—the dirt and foreign matter must be removed to permit the refrigerant to pass through the filter.

## Copper Tubing

Many smaller air-conditioning units make use of copper tubing for the liquid and suction lines. Since copper is not structurally as strong as iron or steel, it is relatively easy to collapse the line by an accidental blow, thus preventing the circulation of the refrigerant. When this happens, a new line or section of the line must be installed.

## Expansion Valves

Expansion valves must be properly adjusted and in perfect working order to maintain the correct pressure in the low side of the line. In many cases, oil from the compressor crankcase enters and remains in the evaporator, occupying the space intended for the refrigerant. Naturally, this reduces the refrigerating effect, and service problems develop. The oil must be drained from the evaporator coil and the compressor crankcase inspected to make sure the oil level is correct.

## Compressors

Compressors will start developing trouble after long use as the pistons and rings, as well as the other components, become worn. This wear allows some of the high-pressure gas to leak by the pistons and rings so that proper compression cannot be obtained, and the compressor becomes inefficient and unable to take care of the load. New rings, pistons, and connecting rods usually are all that are required to bring the compressor back to proper performance. Suction and discharge valves may stick, crack, or fail entirely. Sometimes, a slight warpage or piece of dirt under them results in improper operation. Seals around the crankshaft may leak and require repacking or replacement.

## Condensers

The condensers may become dirty and inefficient, thereby resulting in high pressure, loss of efficiency, and an increase in the power requirement.

## Ducts

Some air-conditioning ducts are provided with insulation on the inside, which may tear loose and flap, resulting in noise.

## Water Jets

Water jets or spray heads (if used) may become clogged or worn. Automatic water valves may get out of adjustment, and the flow of water may become restricted because of clogging of the water strainer.

# SERVICING CHECKS

The proper diagnosis of air-conditioning system service problems takes patience and diligence. The important steps in checking a system are as follows:

1. Determine the refrigerant used in the system. This is an important factor, since each refrigerant has its own operating characteristics, such as pressure-temperature differences.
2. Install a gauge-test set. If the unit is a large one, it will have a set of gauges as part of the equipment. The gauge test will indicate the condition of the frigerant by checking the pressure-temperature relations. Put a thermometer on the evaporator coil, near the expansion valve, and obtain a reading. This temperature reading, along with a back-pressure reading, will indicate the refrigerant condition. Then check for improper expansion-coil pressure.
3. A low-pressure reading may be caused by a shortage of refrigerant, the presence of ice or water in the adjustment side of the expansion valve, moisture in the refrigerant system, plugged screens, partly closed liquid shut-off or service valves, excessive charge, or air in the system. Improper valve settings, blocked air circulation over the condenser, cooling air passing over the condenser at too high a temperature, reversed rotation of the motor, bent fan blades, or a clogged condenser are indicated by high head pressures.
4. Frost or a sweat line on the coil should be noted. A coil is only active up to its frost or sweat line. All lengths of tubing beyond the frost or sweat line are inactive, since they do not receive liquid refrigerant. For greatest efficiency, the entire coil must have frost or sweat, and if the liquid line is obstructed, the screen plugged, or the expansion improperly set, the proper amount of liquid refrigerant will not enter the evaporator, and only a limited portion of the coil will frost or sweat, depending on the type of liquid refrigerant used.
5. An increase or oversupply of refrigerant will cause the entire coil, and possibly the suction line, to frost or sweat. This may also be caused by a leaky needle, improper adjustment, ice in the adjustment side of the expansion valve, or fused thermo-

stat contacts. Most complaints relating to excessive or insufficient frosting are due to weather conditions.

6. Improper operation will occur if refrigerant has escaped from the system. The first thing to do when leakage has occurred is to detect and repair the leak. Indications of refrigerant shortage include hissing at the expansion valve, a warm or hot liquid line, little or no frost on the expansion valve or coil, continuous operation, low head pressure, and bubbles in the signt glass if it is inserted into the liquid line and used to test for refrigerant shortage.

7. Bubbles may form in the sight glass if the head pressure is under 120 lbs. with F-12, and under 100 lbs. with methyl chloride. An excellent indication of leakage is the presence of oil on a joint or fitting; methyl chloride and F-12 dissolve oils, and when a leak occurs, the escaped refrigerant evaporates in the atmosphere, leaving the oil behind. Refrigerant must be added to system having a refrigerant shortage until the bubbles in the sight glass cease, or until the hissing sound at the expansion valve is eliminated.

8. Check for improper installation. Compressor units and low sides must be level. Tubes and fittings forming the liquid and suction lines must be the proper size. Baffles, ducts, and eliminators must be properly located and not obstructed. Ducts must be insulated properly, or short-circuiting of air currents may take place, causing the ducts to sweat. Liquid and suction lines must be checked for pinches, sharp or flattened bends, and obstructions. Lines should not be run along the ceiling in a hot room, and the lines should not run adjacent to any active hot-water or steam pipes.

9. The location and installation of the thermostat must be checked. The thermostatic-switch bulb should be installed in a location where average temperature conditions exist. The thermostatic bulb, or switch itself (if self-contained), should not be placed where an inrush of warm air, such as that caused by the opening of doors or windows, will cause the mechanism to cut in prematurely. Use thermostatic control switches with a minimum of tubing—gas will condense in long runs of tubing, and the condensate will produce erratic operation. If a thermostatic bulb is used to control a liquid

bath or brine, the bulb should be housed in a drywell or cavity, which should be located where average temperature conditions will be stable.

10. The condition of the thermostatic bulb should be checked. Apply heat gently by means of a cloth saturated with hot water, or else hold the bulb in the hand. If the contacts do not close after applying heat, the thermostatic element is discharged. Dirty or oxidized contacts can also cause defective operation.

11. Starter fuses should be checked and, if blown, replaced with fuses of the proper size. Determine the cause of the trouble—shortage of oil, air in the system, overcharge, misalignment, high back-pressure, lack of air or water over the condenser or motor—any of which may cause the fuses to blow. If the trouble is not corrected, serious damage can result in a relatively short time.

12. Filters and/or screens must be checked. A clogged screen or filter, or a pinched or clogged liquid line, will produce the same trouble as a leaky or stuck expansion valve, depending on the degree of the obstruction.

13. A leaky or stuck expansion-valve needle will be indicated by a low pressure in the evaporator side and continuous operation of the unit. Sometimes a low-pressure control is wired in series with the thermostatic control and motor. When this is done, the low-pressure control will cut out, and the unit will remain inoperative.

## COMMON TROUBLES IN AUTOMATIC EXPANSION-VALVE SYSTEMS WITH THERMOSTATIC CONTROL

### Refrigerant Shortage

1. Warm or hot liquid line.
2. Hissing sound at expansion valve.
3. Low head or condensing pressure.
4. Evaporator not entirely chilled.
5. Poor refrigeration.

6. Low-side pressure may be high if only gas is entering the evaporator.

## Poor Refrigeration

1. Heavy cost of frost or ice on the evaporator.
2. Refrigerant shortage.
3. Thermostat not adjusted properly.
4. Thermostat defective.
5. Thermostat shifted and not level.
6. Thermostat shielded by covering.
7. Thermostat in draft.
8. Stuck expansion valve.
9. Expansion valve set too low; allows only a portion of the evaporator to be effective.
10. Liquid line pinched or restricted.
11. Suction line pinched or restricted.
12. Compressor valves defective, broken, or sticking.
13. Strainer on liquid or suction line clogged.
14. Partially closed liquid- or suction-line valves.
15. Ice or moisture in adjustment side of expansion valve.
16. Ice freezing in seat of expansion valve.
17. High head pressure.
18. Compressor losing efficiency through wear.
19. Compressor may be too small.

## Compressor Discharge Valve Defective

1. Low head pressure.
2. Poor refrigeration.
3. Compressor excessively warm.
4. If compressor is stopped, pressure will equalize.

## Expansion-Valve Needle Stuck Open

1. Poor refrigeration.
2. High head pressure if stuck partially open.
3. Low head pressure and high back-pressure if stuck fully open.
4. Frosted or sweating suction line.

5. Hissing sound at expansion valve.
6. Impossible to adjust for higher or lower back-pressure.
7. On methyl chloride and F-12 units, this may be caused by moisture freezing at seat of expansion valve.
8. Improper oil, freezing at seat.
9. Improper oil and high compressor temperature may result in carbonization, which may build up at the expansion valve, especially if filters are defective.

## Expansion-Valve Needle Stuck Closed

1. No refrigeration if shut tight.
2. Little refrigeration if stuck partially shut.
3. Evaporator will be pumped down so that low side will show an unduly low pressure.
4. A pinched liquid line, plugged filter, or closed hand valve will give the same symptoms.
5. On methyl chloride and F-12 units, this may be caused by moisture freezing in the expansion valves.

## High Head Pressure

1. Air in system.
2. Excessive refrigerant charge.
3. Air or water going through condenser at too high a temperature.
4. Air circulation over condenser blocked, if the unit is an air-cooled type.
5. If the unit is of the water-cooled type, the water may be turned off or restricted.
6. Rotation of motor reversed.
7. If a higher setting is used on the expansion valve, the head pressure will be higher, and vice versa.

## Unable to Adjust Valve

1. Refrigeration shortage.
2. Compressor valve broken or defective.
3. Partially plugged screens or filters.
4. Liquid line pinched almost closed.

5. Stoppage in fitting or restriction in liquid line.
6. Needle or seat eroded and leaky.
7. Oil-logged coil.
8. Ice in adjustment side of expansion.

### Low Head Pressure

1. Refrigerant shortage.
2. Worn pistons.
3. Head or clearance gasket too thick.
4. Suction valve worn, split, or stuck.
5. Low setting on expansion valve.
6. Gasket between cylinders blown.

### Suction Line and Drier Coil Frosted or Sweating

1. Expansion valve stuck open or leaky.
2. Needle seat eroded or corroded.
3. Valve set at too high a back-pressure.
4. Ice or moisture in adjustment side of expansion valve.
5. Thermostat out of order or poorly located.
6. Fan not operating so that air is not blown over coils.
7. No water, or water pump not operating to pass water over evaporator.

## COMMON TROUBLES IN THERMOSTATIC EXPANSION-VALVE SYSTEMS WITH LOW-SIDE OR THERMOSTATIC CONTROL

### Refrigerant Shortage

1. Continuous operation.
2. Low head pressure.
3. Poor refrigeration.
4. Warm or hot liquid line.
5. Evaporator coils not chilled throughout entire length.
6. Hissing at expansion line.

## Poor Refrigeration

1. Heavy coating of ice or frost on evaporator coils.
2. Expansion valve set too high.
3. Refrigerant shortage.
4. Thermostat bulb placed where there is little change in coil temperature.
5. Thermostat bulb placed where it is in a cold pocket and not affected by average conditions.
6. Expansion valve set too low.
7. Pigtail of expansion valve improperly placed so that maximum coil surface is not used.
8. Compressor valves defective, broken, or sticking.
9. Liquid line pinched.
10. Suction line pinched.
11. Strainer clogged.
12. Suction line too small for job.
13. Partially closed liquid- or suction-line hand valves.
14. Compressor too small for job.
15. Moisture in methyl chloride or F-12 refrigerant.

## Compressor Discharge Valve Defective

1. Low head pressure.
2. Poor refrigeration.
3. When compressor is stopped, pressures equalize.

## Expansion-Valve Needle Stuck Open

1. Continuous operation.
2. Poor refrigeration.
3. High head pressure.
4. Hissing sound at expansion valve.
5. Moisture in methyl chloride or F-12 refrigerant.

## Low Head Pressure

1. Shortage of refrigerant.
2. Worn pistons in compressor.
3. Warped, split, or stuck discharge valve.
4. Suction valve warped, split, or stuck.

5. Expansion-valve needle stuck wide open.
6. Gasket between cylinder blown.
7. Thermostatic bulb discharged.

## Suction Line and Drier Coil Sweating or Frosted

1. Expansion-valve needle stuck open.
2. Expansion-valve needle or seat eroded and leaky.
3. Expansion valve set too high above cutout point.
4. Control-switch points fused together.
5. Low-side control switch locked in operating position.

## Expansion-Valve Needle Stuck Partially Closed

1. Little or no refrigeration.
2. The high-pressure safety cut-out may trip.
3. Evaporator will be pumped down and show a low pressure at below cut-in temperature.
4. If the liquid line is plugged, the thermostatic bulb discharged, or the capillary tube pinched, the result will be the same as a needle stuck shut.

## High Head Pressure

1. The high-pressure safety cut-out may cause system to be stopped.
2. Air in the system.
3. Excessive refrigerant charge.
4. Air or water passing over condenser at too high a temperature.
5. If unit is water-cooled, flow may be restricted or turned off.
6. If a high setting is used on the expansion valve (resulting in high back-pressure), the head pressure will be higher than if the suction pressure were low.
7. Rotation of fan for cooling condenser reversed.
8. Fan blades bent or air passing over condenser restricted.

## Expansion Valve Cannot Be Adjusted

1. Oil-logged evaporator.
2. Shortage of refrigerant.

3. Compressor valve broken or stuck.
4. Partially plugged screen in filter.
5. Liquid line pinched.
6. Stoppage in fitting or restriction in liquid line.
7. Stop-valve seat dropped and sealing open.
8. Charge lost in valve bulb.
9. Valve bulb loosened in its cradle by frost action; not making proper contact.
10. Valve covered; not open to atmospheric conditions.
11. Valve in too cold a location.

## REFRIGERANT PRESSURES

Table 10-4 lists twelve of the common refrigerants that can be encountered in almost any unit. The column at the left in the table is temperature in degrees Fahrenheit. Each refrigerant is listed in separate columns, and the pressures given are listed in *pounds of pressure per square inch (gauge) (psig)*, or in *inches of mercury* if the pressures are below zero pounds gauge pressure. The temperatures are given in 5° increments within the range of normal practice. The very low temperatures are seldom employed, but have been included for convenience. If the desired temperature lies between any two of the given temperatures, a little arithmetic can be used to determine the corresponding pressure with sufficient accuracy to set the valves or controls. To determine the vaporization pressure in the evaporator, the temperature of the liquid must be decided on.

For convenience, let us select ammonia as the refrigerant and decide to have the ammonia boil at 10° above zero F. With an ammonia system, the refrigerating plant will probably be located in the basement or on the roof and will be used to cool brine. The cooled brine will be pumped to the cooling coils at about 25°F. First, run down the temperature column until the 10° line is found, then follow the line horizontally to the right, and under the heading "Ammonia," read the pressure of 23.8 psig. The head of condensing pressures can then be checked. This is done by taking the temperature of the outlet water, if the refrigerating unit is water-cooled. When we assume that the water is leaving the condenser at a temperature of 90°F, a reference to the table indicates that the head pressure should be in the vicinity of 166 psig. Pressures will usually be a few pounds higher in actual practice.

If air-cooled apparatus is encountered, the temperature of the air passing over the condenser must first be determined. Let us assume that the air passing over the condenser is at a temperature of 70°F. In Table 10-5, it will be observed that a temperature column is also given, which is the temperature of the liquid in the evaporator. In our particular case, the ammonia is evaporating at a temperature of 10°F. Therefore, find this specific temperature in the left-hand column and then continue horizontally from that point to the column headed 70°F. At this intersection, the figure 35 is given. This figure (35) is added to the temperature of the air (70), which gives a total of 105°F. Turn to Table 10-4 and find the 105° line, which is the pressure corresponding to the temperature that exists in the condenser. Table 10-4 gives the temperature at which boiling or vaporization of each particular refrigerant takes place; the pressure at which this occurs is listed under the refrigerant type.

If the refrigerant is ammonia at 15.7 psig, the liquid will boil at 0°F. If it is desired to have a somewhat colder liquid, alter the pressure. This is accomplished in the refrigerating apparatus by adjusting the temperature control, if a thermostat is the control feature (such as with a low-side float), or by adjusting the thermostat or pressurestat in either a high-side or low-side float system. The continued operation will lower the evaporator temperature and the temperature of the room in which the evaporator is located. With an expansion-valve system, if a thermostat is employed to stop and start the system, the thermostat can be adjusted for a higher or lower temperature. Before and after the adjustment, it is important that the frost line on the evaporator be checked.

If the expansion valve is opened to admit more refrigerant so that the pressure is raised and its boiling temperature increased, the suction line may frost back. Also, if the expansion valve is closed so that less refrigerant enters the evaporator, the compressor will be unable to hold the evaporator at a lower pressure and a lower temperature. When this is done, only a portion of the coil may be effective. A coil must contain liquid refrigerant to be able

355

## Table 10-4. Pressure in Psig or Inches of Vacuum Corresponding to Temperature in °F for Various Common Refrigerants

| Temp., °F | Ammonia NH₃ | Sulfur dioxide SO₂ | Methyl chloride CH₃Cl | Ethane C₂H₆ | Propane C₃H₈ | Ethyl chloride C₂H₅Cl | Carbon dioxide CO₂ | Butane C₄H₁₀ | Isobutane C₄H₁₀ | Freon F-12 CCl₂F₂ | Carrene CH₂Cl₂ | Methyl formate C₂H₄O₂ |
|---|---|---|---|---|---|---|---|---|---|---|---|---|
| -40 | 8.7" | 23.5" | 15.7" | 99.8# | 1.5# | | 131.1# | | 11.0" | | | |
| -35 | 5.4" | 22.4" | 14.4" | 109.8# | 3.4# | | 156.3# | | 8.4" | | | |
| -30 | 1.6" | 21.1" | 11.6" | 120.3# | 5.6# | | 163.1# | | 5.5" | | | |
| -25 | 1.3# | 19.6" | 9.2" | 132.0# | 8.0# | | 176.3# | | 2.3" | | | |
| -20 | 3.6# | 17.9" | 6.1" | 144.8# | 10.7# | 25.3" | 205.8# | | 14.6" | 0.5# | | |
| -15 | 6.2# | 16.1" | 2.3" | 157 # | 13.6# | 24.5" | 225.8# | | 13.0" | 2.4# | | |
| -10 | 9.0# | 13.9" | 0.2# | 172 # | 16.7# | 23.6" | 247.0# | | 11.0" | 4.5# | 28.1" | |
| -5 | 12.2# | 11.5" | 2.0# | 187 # | 20.0# | 22.6" | 269.7" | | 8.8" | 6.8# | 27.8" | |
| 0 | 15.7# | 8.8" | 3.8# | 204 # | 23.5# | 21.5" | 293.9# | 15.0" | 6.3" | 9.2# | 27.5" | 26.5" |
| +5 | 19.6# | 5.8" | 6.2# | 221 # | 27.4# | 20.3" | 319.7# | 12.2" | 3.3" | 11.9# | 27.1" | 25.9" |
| +10 | 23.8# | 2.6" | 8.6# | 239 # | 31.4# | 18.9" | 347.1# | 11.1" | 0.2" | 14.7# | 26.7" | 25.4" |
| +15 | 28.4# | 0.5" | 11.2# | 257 # | 35.9# | 17.4" | 376.3# | 8.8" | 1.6# | 17.7# | 26.2" | 24.7" |
| +20 | 33.5# | 2.4# | 13.6# | 227 # | 40.8# | 15.8" | 407.3# | 6.3" | 3.5# | 21.1# | 25.6" | 24.0" |
| +25 | 39.0# | 4.6# | 17.2# | 292 # | 46.2# | 14.0" | 440.1# | 3.6" | 5.5# | 24.6# | 24.9" | 23.1" |
| +30 | 45.0# | 7.0# | 20.3# | 320 # | 51.6# | 12.2" | 474.9# | 0.6" | 7.6# | 28.5# | 24.3" | 22.3" |
| +35 | 51.6# | 9.6# | 24.0# | 343 # | 57.3# | 10.1" | 511.7# | 1.3# | 9.9# | 32.6# | 23.5" | 21.1" |

## Table 10-4. Pressure in Psig or Inches of Vacuum Corresponding to Temperature in °F for Various Common Refrigerants (Cont'd)

| °F | | | | | | | | | | | |
|---|---|---|---|---|---|---|---|---|---|---|---|
| +40 | 58.6# | 12.4# | 28.1# | 368 # | 63.3# | 550.7# | 3.0# | 12.2# | 37.0# | 22.6" | 20.0" |
| +45 | 66.3# | 15.5# | 32.2# | 390 # | 69.9# | 591.8# | 4.9# | 14.8# | 41.7# | 21.7" | 18.7" |
| +50 | 74.5# | 18.8# | 36.3# | 413 # | 77.1# | 635.3# | 6.9# | 17.8# | 46.7# | 20.7" | 17.3" |
| +55 | 83.4# | 22.4# | 41.7# | 438 # | 84.6# | 681.2# | 9.1# | 20.8# | 52.0# | 19.5" | 15.7" |
| +60 | 92.9# | 26.2# | 46.3# | 466 # | 92.4# | 729.5# | 11.6# | 24.0# | 57.7# | 18.2" | 14.0" |
| +65 | 103.1# | 30.4# | 53.6# | 496 # | 100.7# | 780.4# | 14.2# | 27.5# | 63.7# | 16.7" | 11.9" |
| +70 | 114.1# | 34.9# | 57.8# | 528 # | 109.3# | 834.0# | 16.9# | 31.1# | 70.1# | 15.1" | 9.8" |
| +75 | 125.8# | 39.8# | 64.4# | 569 # | 118.5# | 890.4# | 19.8# | 35.0# | 76.9# | 13.4" | 7.3" |
| +80 | 138.3# | 45.0# | 72.3# | 610 # | 128.1# | 949.6# | 22.9# | 39.2# | 84.1# | 11.5" | 4.9" |
| +85 | 151.7# | 50.9# | 79.4# | 657 # | 138.4# | 1011.3# | 26.2# | 43.9# | 91.7# | 8.4" | 2.4" |
| +90 | 165.9# | 56.5# | 87.3# | 693 # | 149 # | | 29.8# | 48.6# | 99.6# | 7.3" | 0.5# |
| +95 | 181.1# | 62.9# | 95.6# | | 160 # | | 33.2# | 53.7# | 108.1# | 5.0" | 2.1# |
| +100 | 197.2# | 69.8# | 102.3# | | 172 # | | 37.5# | 59.0# | 116.9# | 2.4" | 3.8# |
| +105 | 214.2# | 77.1# | 113.4# | | 185 # | | 41.7# | 64.6# | 126.2# | 0.19# | 5.8# |
| +110 | 232.2# | 85.1# | 118.3# | | 197 # | | 46.1# | 70.4# | 136.0# | 1.6# | 7.7# |
| +115 | 251.5# | 93.5# | 128.6# | | 207.6# | | | 76.7# | 146.5# | 3.1# | 10.4# |
| +120 | 271.7# | 106.4# | 139.3# | | 218.3# | | | 84.3# | 157.1# | 4.7# | 13.1# |
| +125 | 293.1# | 111.9# | 150.3# | | 232.3# | | | 90.1# | 168.6# | 6.6# | 15.7# |
| +130 | 315.6# | 121.9# | 161.3# | | 246.3# | | | 97.3# | 180.2# | 8.4# | 18.2# |

" = inches of mercury
# = psig

to refrigerate. Therefore, if only part of the coil is frosted, the other part may just as well be taken out, since it is doing no work. The best plan is to maintain the highest pressure possible so that the compressor will work on the densest vapor without having the suction line frost back.

**Table 10-5. Factors to be Added to Initial Temperatures of Coolant to Determine Condenser Temperatures**

| Evap. | Initial Coolant Temp. °F | | | | |
|---|---|---|---|---|---|
| °F | 60° | 70° | 80° | 90° | 100° |
| −30 | 15 | 15 | 15 | 10 | 10 |
| −25 | 15 | 15 | 15 | 15 | 10 |
| −20 | 20 | 20 | 15 | 15 | 15 |
| −15 | 20 | 20 | 20 | 15 | 15 |
| −10 | 20 | 20 | 20 | 15 | 15 |
| − 5 | 20 | 20 | 20 | 15 | 15 |
| 0 | 25 | 25 | 20 | 20 | 15 |
| + 5 | 30 | 30 | 30 | 25 | 20 |
| +10 | 35 | 35 | 30 | 25 | 20 |
| +15 | 40 | 35 | 30 | 25 | 20 |
| +20 | 40 | 35 | 30 | 25 | 25 |
| +25 | 45 | 40 | 35 | 30 | 30 |
| +30 | 50 | 45 | 40 | 35 | 35 |
| +35 | 50 | 50 | 45 | 45 | 40 |

It may be desirable, however, to make a change in the suction pressure so that the temperature of the coils can be altered. Thus, with ammonia, if it is desired that the coil temperature is to be 20°F, the expansion valve must be adjusted to maintain a constant pressure of 33.5 psig in the evaporator. If a 40° temperature is desired, a 58.6 psig reading on the low side would have to be maintained. If a 40° temperature were required in the evaporator of a methyl-chloride system, the expansion valve would have to be adjusted to hold a constant pressure of 28.1 psig.

Table 10-4 lists the temperature of the liquid refrigerant boiling at the constant pressure given. One factor cannot be varied without varying the other. The only way to obtain another boiling temperature is to adjust the expansion valve to maintain the pressure that corresponds to the temperature desired. If the liquid is contained in a vessel or cylinder, and the vessel is closed (such as in a supply drum), it will be observed that the pressure in such a vessel will be from 1 to 5 pounds greater than the temperature of the drum. This is due to the fact that pressure has built up and prohibits further evaporation. If a system contains an unknown refrigerant, such as one without an odor, the kind may be determined by taking a pressure reading of the refrigerant at rest; that is, while the unit is not operating. That pressure that corresponds nearest to the reading will be that particular refrigerant tested.

Table 10-4 is based on determining, from a pressure reading of the low side or evaporator, just what the temperature of the boiling liquid may be. With a methyl-chloride system, if the low-pressure gauge reads 2 psig, the temperature corresponding to this pressure is given as −5°F in the table.

With the temperature of the evaporator known ( −5°F), refer to Table 10-5. The first column in the table on the left-hand side is the evaporator temperature. The other columns refer to the temperature of the water or air passing over the condenser. Assume, for instance, that the room temperature is 70°F, and that the air is passing over the condenser at this temperature. Since the evaporator temperature is known, find this temperature in the first column and then find the proper column for the coolant temperature (70°F). At the intersection under these conditions, the figure 20 is given. This is added to the initial temperature, thus: 20° + 70° = 90°. Take this figure (90°) and refer to Table 10-4. Run down the temperature column to 90° and then across to the methyl-chloride column. It will be found that a head or condensing pressure of about 87.3 psig can be expected. If the actual pressure reading varies from this, check the condition of the apparatus.

It is essential to obtain a pressure reading on the low or evaporator side so that the temperature of the refrigerant can be determined. If the reading indicates too high a temperature, the expansion valve will have to be adjusted to maintain the proper pressure and, of course, the proper evaporating temperature. Head pressures are important; excessive pressures indicate something out of the ordinary.

# AIR-CONDITIONING
# SYSTEM TROUBLE CHART

## Compressor Will Not Start

| Possible Cause | Possible Remedy |
|---|---|
| Thermostat setting too high. | Reset thermostat below room temperature. |
| High head pressure. | Reset starter overload and determine cause of high head pressure. |
| Defective pressure switch. | Repair or replace pressure switch. |
| Loss of refrigerant charge. | Check system for leaks. |
| Compressor frozen. | Replace compressor. |

## Compressor Short-Cycles

| Possible Cause | Possible Remedy |
|---|---|
| Defective thermostat. | Replace thermostat. |
| Incorrect setting on low side of pressure switch. | Reset low-pressure switch differential. |
| Low refrigerant charge. | Check system for leaks; repair and add refrigerant. |
| Defective overload. | Replace overload. |
| Dirty or iced evaporator. | Clean or defrost evaporator. |
| Evaporator blower and motor belts slipping. | Tighten or replace belts. |
| Dirty or plugged filters. | Clean or replace air filters. |

## Compressor Runs Continuously

| Possible Cause | Possible Remedy |
|---|---|
| Excessive load. | Check for excessive outside air infiltration and excessive source of moisture. |

## Compressor Runs Continuously (Cont'd)

| Possible Cause | Possible Remedy |
| --- | --- |
| Air or noncondensable gases in the system. | Purge system. |
| Dirty condenser. | Clean condenser. |
| Condenser blower and motor belts slipping. | Tighten or replace belts. |
| Thermostat setting too low. | Reset thermostat. |
| Low refrigerant charge. | Check system for leaks; repair and add refrigerant. |
| Overcharge of refrigerant. | Purge and remove excess refrigerant. |
| Compressor valves leaking. | Replace compressor. |
| Expansion valve or strainer plugged. | Clean expansion valve or strainer. |

## System Short of Capacity

| Possible Cause | Possible Remedy |
| --- | --- |
| Low refrigerant charge. | Check system for leaks; repair and add refrigerant. |
| Incorrect superheat setting of expansion valve. | Adjust superheat to 10°F. |
| Defective expansion valve. | Repair or replace valve. |
| Air or noncondensable gases in the system. | Purge system. |
| Condenser blower and motor belts slipping. | Tighten or replace belts. |
| Overcharge of refrigerant. | Purge excess refrigerant. |
| Compressor valves leaking. | Replace compressor valves. |
| Expansion valve or strainer plugged. | Clean valve or strainer. |
| Condenser air short-circuiting. | Remove obstructions or causes of short. |

## Head Pressure Too High

| Possible Cause | Possible Remedy |
| --- | --- |
| Overcharge of refrigerant. | Purge excess refrigerant. |
| Air or noncondensable gases in system. | Purge system. |
| Dirty condenser. | Clean condenser. |
| Condenser blower and motor belts slipping. | Tighten or replace belts. |
| Condenser air short-circuiting. | Remove obstructions or causes of short-circuiting air. |

## Head Pressure Too Low

| Possible Cause | Possible Remedy |
| --- | --- |
| Low refrigerant charge. | Check system for leaks; repair and add refrigerant. |
| Compressor valves leaking. | Replace compressor valves. |

## Suction Pressure Too High

| Possible Cause | Possible Remedy |
| --- | --- |
| Excessive load on system. | Remove conditions causing excessive load. |
| Expansion valve is stuck in "open" position. | Repair or replace expansion valve. |
| Incorrect superheat setting of expansion valve. | Adjust superheat setting to 10°F. |

## Suction Pressure Too Low

| Possible Cause | Possible Remedy |
| --- | --- |
| Low refrigerant charge. | Check system for leaks; repair and add refrigerant. |

| Expansion valve or strainer plugged. | Clean expansion valve or strainer. |
| Incorrect superheat setting of expansion valve. | Adjust superheat setting to 10°. |
| Evaporator air volume low. | Increase air over evaporator. |
| Stratification of cool air in conditioned area. | Increases air velocity through grilles. |

### Compressor Is Noisy

| Possible Cause | Possible Remedy |
| --- | --- |
| Worn or scored compressor bearings. | Replace compressor. |
| Expansion valve is stuck in "Open" position or defective. | Repair or replace expansion valve. |
| Overcharge of refrigerant or air in system. | Purge system. |

| Possible Cause | Possible Remedy |
| --- | --- |
| Liquid refrigerant flooding back to compressor. | Repair or replace expansion valve. |
| Shipping or hold-down bolts not loosened. | Loosen compressor hold-down bolts so compressor is freely floating in mountings. |

### Compressor and Condenser Fan Motor Will Not Start

| Possible Cause | Possible Remedy |
| --- | --- |
| Power failure. | Check electrical wiring back to fuse box. |
| Fuse blown. | Replace blown or defective fuse. |
| Thermostat setting too high. | Reduce temperature setting of room thermostat. |
| Defective thermostat. | Replace or repair thermostat. |
| Faulty wiring. | Check wiring and make necessary repairs. |

363

Defective controls.

Check and replace defective controls.

Low voltage.

Reset and check for cause of tripping.

Defective dual-pressure control.

Replace the control.

## Compressor Will Not Start, but the Condenser Fan Motor Runs

| Possible Cause | Possible Remedy |
| --- | --- |
| Faulty wiring to compressor. | Check compressor wiring and repair. |
| Defective compressor motor. | Replace the compressor. |
| Defective compressor overload (single phase only). | Replace overload. |
| Defective starting capacitor (single phase only). | Replace capacitor. |

## Condenser Fan Motor Will Not Start, but Compressor Runs

| Possible Cause | Possible Remedy |
| --- | --- |
| Faulty wiring to fan motor. | Check fan-motor wiring and repair. |
| Defective fan motor. | Replace fan motor. |

## Condenser Fan Motor Runs, but the Compressor Hums and Will Not Start

| Possible Cause | Possible Remedy |
| --- | --- |
| Low voltage. | Check line voltage. Determine the location of the voltage drop. |
| Faulty wiring. | Check wiring and make necessary repairs. |

| Defective compressor. | Replace compressor. |
| High head pressure. | Check head pressure and complete operation of system to remove the cause of the high pressure condition. |
| Failure of one phase (three-phase units only). | Check fuses and wiring. |
| Defective start capacitor (single phase only). | Replace capacitor. |
| Defective potential relay (single phase only). | Replace relay. |

## Compressor Starts, but Cycles on Overload

| Possible Cause | Possible Remedy |
|---|---|
| Low voltage. | Check line voltage. Determine the location of the voltage drop. |
| Faulty wiring. | Check wiring and make necessary repairs. |
| Defective running capacitor (single phase only). | Replace capacitor. |
| Defective overload. | Replace overload. |
| Unbalanced line (three-phase only). | Check wiring; call power company. |

## Evaporator Fan Motor Will Not Start

| Possible Cause | Possible Remedy |
|---|---|
| Power failure. | Check electrical wiring back to fuse box. |

365

# Index

# The Audel®
# Mail Order
# Bookstore

Here's an opportunity to order the valuable books you may have missed before and to build your own personal, comprehensive library of Audel books. You can choose from an extensive selection of technical guides and reference books. They will provide access to the same sources the experts use, put all the answers at your fingertips, and give you the know-how to complete even the most complicated building or repairing job, in the same professional way.

## Each volume:

- **Fully illustrated**
- **Packed with up-to-date facts and figures**
- **Completely indexed for easy reference**

## APPLIANCES
### HOME APPLIANCE SERVICING, 4th Edition
A practical book for electric & gas servicemen, mechanics & dealers. Covers the principles, servicing, and repairing of home appliances. 592 pages; 5½ × 8¼; hardbound. **Price: $15.95**

### REFRIGERATION: HOME AND COMMERCIAL
Covers the whole realm of refrigeration equipment from fractional-horsepower water coolers through domestic refrigerators to multiton commercial installations. 656 pages; 5½ × 8¼; hardbound. **Price: $16.95**

### AIR CONDITIONING: HOME AND COMMERCIAL
A concise collection of basic information, tables, and charts for those interested in understanding troubleshooting, and repairing home air-conditioners and commercial installations. 464 pages; 5½ × 8¼; hardbound. **Price: $14.95**

### OIL BURNERS, 4th Edition
Provides complete information on all types of oil burners and associated equipment. Discusses burners—blowers—ignition transformers—electrodes—nozzles—fuel pumps—filters—controls. Installation and maintenance are stressed. 320 pages; 5½ × 8¼; hardbound. **Price: $12.95**

## AUTOMOTIVE
### AUTOMOBILE REPAIR GUIDE, 4th Edition
A practical reference for auto mechanics, servicemen, trainees, and owners. Explains theory, construction, and servicing of modern domestic motorcars. 800 pages; 5½ × 8¼; hardbound. **Price: $14.95**

---

*Use the order coupon on the back of this book.*
All prices are subject to change without notice.

## AUTOMOTIVE AIR CONDITIONING

You can easily perform most all service procedures you've been paying for in the past. This book covers the systems built by the major manufacturers, even after-market installations. Contents: introduction—refrigerant—tools—air conditioning circuit—general service procedures—electrical systems—the cooling systems—system diagnosis—electrical diagnosis—troubleshooting. 232 pages; 5½ × 8¼; softcover. **Price: $7.95**

## DIESEL ENGINE MANUAL, 4th Edition

A practical guide covering the theory, operation and maintenance of modern diesel engines. Explains diesel principles—valves—timing—fuel pumps—pistons and rings—cylinders—lubrication—cooling system—fuel oil and more. 480 pages; 5½ × 8¼; hardbound. **Price: $12.95**

## GAS ENGINE MANUAL, 2nd Edition

A completely practical book covering the construction, operation, and repair of all types of modern gas engines. 400 pages; 5½ × 8¼; hardbound. **Price: $9.95**

## SMALL GASOLINE ENGINES

A new manual providing practical and theoretical information for those who want to maintain and overhaul two- and four-cycle engines such as lawn mowers, edgers, snowblowers, outboard motors, electrical generators, and other equipment using engines up to 10 horsepower. 624 pp; 5½ × 8¼; hardbound. **Price: $15.95**

## TRUCK GUIDE—3 Vols.

Three all-new volumes provide a primary source of practical information on truck operation and maintenance. Covers everything from basic principles (truck classification, construction components, and capabilities) to troubleshooting and repair. 1584 pages; 5½ × 8¼; hardbound.
**Price: $41.85**
> **Volume 1**
> ENGINES: **$14.95**
> **Volume 2**
> ENGINE AUXILIARY SYSTEMS: **$14.95**
> **Volume 3**
> TRANSMISSIONS, STEERING AND BRAKES: **$14.95**

# BUILDING AND MAINTENANCE
## ANSWERS ON BLUEPRINT READING, 3rd Edition

Covers all types of blueprint reading for mechanics and builders. This book reveals the secret language of blueprints, step by step in easy stages. 312 pages; 5½ × 8¼; hardbound.
**Price: $9.95**

## BUILDING MAINTENANCE, 2nd Edition

Covers all the practical aspects of building maintenance. Painting and decorating; plumbing and pipe fitting; carpentry; heating maintenance; custodial practices and more. (A book for building owners, managers, and maintenance personnel.) 384 pages; 5½ × 8¼; hardbound.
**Price: $9.95**

## COMPLETE BUILDING CONSTRUCTION

At last—a one volume instruction manual to show you how to construct a frame or brick building from the footings to the ridge. Build your own garage, tool shed, other outbuildings—even your own house or place of business. Building construction tells you how to lay out the building and excavation lines on the lot; how to make concrete forms and pour the footings and foundation; how to make concrete slabs, walks, and driveways; how to lay concrete block, brick and tile; how to build your own fireplace and chimney. It's one of the newest Audel books, clearly written by experts in each field and ready to help you every step of the way. 800 pages; 5½ × 8¼; hardbound. **Price: $19.95**

*Use the order coupon on the back of this book.*
All prices are subject to change without notice.

## GARDENING, LANDSCAPING, & GROUNDS MAINTENANCE, 3rd Edition

A comprehensive guide for homeowners and for industrial, municipal, and estate grounds-keepers. Gives information on proper care of annual and perennial flowers; various house plants; green-house design and construction; insect and rodent controls; and more. 416 pages; 5½ × 8¼; hardbound. **Price: $15.95**

## CARPENTERS & BUILDERS LIBRARY, 5th Edition (4 Vols.)

A practical, illustrated trade assistant on modern construction for carpenters, builders, and all woodworkers. Explains in practical, concise language and illustrations all the principles, advances, and shortcuts based on modern practice. How to calculate various jobs. **Price: $39.95**

### Volume 1
Tools, steel square, saw filing, joinery cabinets. 384 pages; 5½ × 8¼; hardbound.
**Price: $10.95**

### Volume 2
Mathematics, plans, specifications, estimates. 304 pages; 5½ × 8¼; hardbound.
**Price: $10.95**

### Volume 3
House and roof framing, layout foundations. 304 pages; 5½ × 8¼; hardbound.
**Price: $10.95**

### Volume 4
Doors, windows, stairs, millwork, painting. 368 pages; 5½ × 8¼; hardbound.
**Price: $10.95**

## HEATING, VENTILATING, AND AIR CONDITIONING LIBRARY (3 Vols.)

This three-volume set covers all types of furnaces, ductwork, air conditioners, heat pumps, radiant heaters, and water heaters, including swimming-pool heating systems. **Price: $41.95**

### Volume 1
Partial Contents: Heating Fundamentals—Insulation Principles—Heating Fuels—Electric Heating System—Furnace Fundamentals—Gas-Fired Furnaces—Oil-Fired Furnaces—Coal-Fired Furnaces —Electric Furnaces. 614 pages; 5½ × 8¼; hardbound. **Price: $14.95**

### Volume 2
Partial Contents: Oil Burners—Gas Burners—Thermostats and Humidistats—Gas and Oil Controls—Pipes, Pipe Fitting, and Piping Details—Valves and Valve Installations. 560 pages; 5½ × 8¼; hardbound. **Price: $14.95**

### Volume 3
Partial Contents: Radiant Heating—Radiators, Convectors, and Unit Heaters—Stoves, Fireplaces, and Chimneys—Water Heaters and Other Appliances—Central Air Conditioning Systems—Humidifiers and Dehumidifiers. 544 pages; 5½ × 8¼; hardbound. **Price: $14.95**

## HOME-MAINTENANCE AND REPAIR: Walls, Ceilings, and Floors

Easy-to-follow instructions for sprucing up and repairing the walls, ceiling, and floors of your home. Covers nail pops, plaster repair, painting, paneling, ceiling and bathroom tile, and sound control. 80 pages; 8½ × 11; softcover. **Price: $6.95**

## HOME PLUMBING HANDBOOK, 3rd Edition

A complete guide to home plumbing repair and installation, 200 pages; 8½ × 11; softcover.
**Price: $8.95**

## MASONS AND BUILDERS LIBRARY, 2nd Edition—2 Vols.

A practical, illustrated trade assistant on modern construction for bricklayers, stonemasons, cement workers, plasterers, and tile setters. Explains all the principles, advances, and shortcuts based on modern practice—including how to figure and calculate various jobs. **Price: $24.90**

### Volume 1
Concrete Block, Tile, Terrazzo. 368 pages; 5½ × 8¼; hardbound. **Price: $12.95**

---

***Use the order coupon on the back of this book.***
All prices are subject to change without notice.

**Volume 2**

Bricklaying, Plastering Rock Masonry, Clay Tile. 384 pages; 5½ × 8¼; hardbound.
**Price: $12.95**

## PAINTING AND DECORATING

This all-inclusive guide to the principles and practice of coating and finishing interior and exterior surfaces is a fundamental sourcebook for the working painter and decorator and an invaluable guide for the serious amateur or building owner. Provides detailed descriptions of materials, pigmenting and mixing procedures, equipment, surface preparation, restoration, repair, and antiquing of all kinds of surfaces. 608 pages; 5½ × 8¼; hardbound. **Price: $18.95**

## PLUMBERS AND PIPE FITTERS LIBRARY, 3rd Edition—3 Vols.

A practical, illustrated trade assistant and reference for master plumbers, journeymen and apprentice pipe fitters, gas fitters and helpers, builders, contractors, and engineers. Explains in simple language, illustrations, diagrams, charts, graphs, and pictures the principles of modern plumbing and pipe-fitting practices. **Price: $32.85**

**Volume 1**

Materials, tools, roughing-in. 320 pages; 5½ × 8¼; hardbound. **Price: $11.95**

**Volume 2**

Welding, heating, air-conditioning. 384 pages; 5½ × 8¼; hardbound. **Price: $11.95**

**Volume 3**

Water supply, drainage, calculations. 272 pages; 5½ × 8¼; hardbound. **Price: $11.95**

## THE PLUMBERS HANDBOOK, 7th Edition

A pocket manual providing reference material for plumbers and/or pipe fitters. General information sections contain data on cast-iron fittings, copper drainage fittings, plastic pipe, and repair of fixtures. 330 pages; 4 × 6 softcover. **Price: $9.95**

## QUESTIONS AND ANSWERS FOR PLUMBERS EXAMINATIONS, 2nd Edition

Answers plumbers' questions about types of fixtures to use, size of pipe to install, design of systems, size and location of septic tank systems, and procedures used in installing material. 256 pages; 5½ × 8¼; softcover. **Price: $8.95**

## TREE CARE MANUAL

The conscientious gardener's guide to healthy, beautiful trees. Covers planting, grafting, fertilizing, pruning, and spraying. Tells how to cope with insects, plant diseases, and environmental damage. 224 pages; 8½ × 11; softcover. **Price: $8.95**

## UPHOLSTERING

Upholstering is explained for the average householder and apprentice upholsterer. From repairing and regluing of the bare frame, to the final sewing or tacking, for antiques and most modern pieces, this book covers it all. 400 pages; 5½ × 8¼; hardbound. **Price: $12.95**

## WOOD FURNITURE: Finishing, Refinishing, Repair

Presents the fundamentals of furniture repair for both veneer and solid wood. Gives complete instructions on refinishing procedures, which includes stripping the old finish, sanding, selecting the finish and using wood fillers. 352 pages; 5½ × 8¼; hardbound. **Price: $9.95**

# ELECTRICITY/ELECTRONICS
## ELECTRICAL LIBRARY

If you are a student of electricity or a practicing electrician, here is a very important and helpful library you should consider owning. You can learn the basics of electricity, study electric motors and wiring diagrams, learn how to interpret the NEC, and prepare for the electrician's examination by using these books.

*Use the order coupon on the back of this book.*

All prices are subject to change without notice.

Electric Motors, 4th Edition. 528 pages; 5½ × 8¼; hardbound. **Price: $12.95**

Guide to the 1984 National Electrical Code. 672 pages; 5½ × 8¼; hardbound.
**Price: $18.95**

House Wiring, 6th Edition. 256 pages; 5½ × 8¼; hardbound. **Price: $12.95**

Practical Electricity, 4th Edition. 496 pages; 5½ × 8¼; hardbound. **Price: $13.95**

Questions and Answers for Electricians Examinations, 8th Edition. 288 pages; 5½ × 8¼;
hardbound. **Price: $12.95**

## ELECTRICAL COURSE FOR APPRENTICES AND JOURNEYMEN, 2nd Edition

A study course for apprentice or journeymen electricians. Covers electrical theory and its applications. 448 pages; 5½ × 8¼; hardbound. **Price: $13.95**

## FRACTIONAL HORSEPOWER ELECTRIC MOTORS

This new book provides guidance in the selection, installation, operation, maintenance, repair, and replacement of the small-to-moderate size electric motors that power home appliances and over 90 percent of industrial equipment. Provides clear explanations and illustrations of both theory and practice. 352 pages; 5½ × 8¼; hardbound. **Price: $15.95**

## TELEVISION SERVICE MANUAL, 5th Edition

Provides the practical information necessary for accurate diagnosis and repair of both black-and-white and color television receivers. 512 pages; 5½ × 8¼; hardbound. **Price: $15.95**

# ENGINEERS/MECHANICS/MACHINISTS
## MACHINISTS LIBRARY, 4th Edition

Covers the modern machine-shop practice. Tells how to set up and operate lathes, screw and milling machines, shapers, drill presses and all other machine tools. A complete reference library.
**Price: $35.85**

### Volume 1
Basic Machine Shop. 352 pages; 5½ × 8¼; hardbound. **Price: $12.95**
### Volume 2
Machine Shop. 480 pages; 5½ × 8¼; hardbound. **Price: $12.95**
### Volume 3
Toolmakers Handy Book. 400 pages; 5½ × 8¼; hardbound. **Price: $12.95**

## MECHANICAL TRADES POCKET MANUAL, 2nd Edition

Provides practical reference material for mechanical tradesmen. This handbook covers methods, tools equipment, procedures, and much more. 256 pages; 4 × 6; softcover. **Price: $10.95**

## MILLWRIGHTS AND MECHANICS GUIDE, 3rd Edition

Practical information on plant installation, operation, and maintenance for millwrights, mechanics, maintenance men, erectors, riggers, foremen, inspectors, and superintendents. 960 pages;
5½ × 8¼; hardbound. **Price: $19.95**

## POWER PLANT ENGINEERS GUIDE, 3rd Edition

The complete steam or diesel power-plant engineer's library. 816 pages; 5½ × 8¼; hardbound.
**Price: $16.95**

## WELDERS GUIDE, 3rd Edition

This new edition is a practical and concise manual on the theory, practical operation and maintenance of all welding machines. Fully covers both electric and oxy-gas welding. 928 pages;
5½ × 8¼; hardbound. **Price: $19.95**

---

*Use the order coupon on the back of this book.*
All prices are subject to change without notice.

## WELDER/FITTERS GUIDE

Provides basic training and instruction for those wishing to become welder/fitters. Step-by-step learning sequences are presented from learning about basic tools and aids used in weldment assembly, through simple work practices, to actual fabrication of weldments. 160 pages; $8^{1}/_{2} \times 11$; softcover. **Price: $7.95**

## FLUID POWER
### PNEUMATICS AND HYDRAULICS, 4th Edition

Fully discusses installation, operation and maintenance of both HYDRAULIC AND PNEUMATIC (air) devices. 496 pages; $5^{1}/_{2} \times 8^{1}/_{4}$; hardbound. **Price: $15.95**

### PUMPS, 4th Edition

A detailed book on all types of pumps from the old-fashioned kitchen variety to the most modern types. Covers construction, application, installation, and troubleshooting. 480 pages; $5^{1}/_{2} \times 8^{1}/_{4}$; hardbound. **Price: $14.95**

### HYDRAULICS FOR OFF-THE-ROAD EQUIPMENT

Everything you need to know from basic hydraulics to troubleshooting hydraulic systems on off-the-road equipment. Heavy-equipment operators, farmers, fork-lift owners and operators, mechanics—all need this practical, fully illustrated manual. 272 pages; $5^{1}/_{2} \times 8^{1}/_{4}$; hardbound. **Price: $8.95**

## HOBBY
### COMPLETE COURSE IN STAINED GLASS

Written by an outstanding artist in the field of stained glass, this book is dedicated to all who love the beauty of the art. Ten complete lessons describe the required materials, how to obtain them, and explicit directions for making several stained glass projects. 80 pages; $8^{1}/_{2} \times 11$; softbound. **Price: $6.95**

---

### *Use the order coupon on the back of this book.*
All prices are subject to change without notice.

## BUILD YOUR OWN AUDEL
## DO-IT-YOURSELF LIBRARY AT HOME!

Use the handy order coupon today to gain the valuable information you need in all the areas that once required a repairman. Save money and have fun while you learn to service your own air conditioner, automobile, and plumbing. Do your own professional carpentry, masonry, and wood furniture refinishing and repair. Build your own security systems. Find out how to repair your TV or Hi-Fi. Learn landscaping, upholstery, electronics and much, much more.